MODERN ARMS
AND
FREE MEN

A DISCUSSION OF
THE ROLE OF SCIENCE IN
PRESERVING DEMOCRACY
BY

VANNEVAR BUSH

SIMON AND SCHUSTER, NEW YORK

1949

PRINTED IN THE UNITED STATES OF AMERICA BY
KINGSPORT PRESS, INC., KINGSPORT, TENNESSEE

To
Henry L. Stimson

ACKNOWLEDGMENTS

While planning and writing this book I turned to a number of friends for criticism of the reasoning and aid on points of fact. Though the book has benefited in many ways through their generosity, responsibility for errors still rests with me, and opinions which are expressed are not to be interpreted as held by those whose courtesy and helpfulness are here acknowledged.

I first wish to record the sympathetic understanding of the late James V. Forrestal, Secretary of National Defense. In the National Military Establishment I am especially indebted to Major General Alfred M. Gruenther, Rear Admiral W. S. Parsons, Major General Kenneth D. Nichols, Major General D. M. Schlatter and William Frye.

Among scientists, Merle A. Tuve, Lloyd V. Berkner, James B. Conant, and Caryl P. Haskins have been especially direct and helpful. I am indebted to Oscar Cox, Elihu Root, Jr., and Eric Hodgins for the encouragement and counsel without which the book could hardly have been completed. Excellent points were brought out by Colonel W. S. Bowen, Redfield Proctor, and Carroll L. Wilson.

Lee Anna Embrey has been of great aid. Among my close associates I am particularly indebted to Paul A. Scherer and Frederick G. Fassett, Jr. The latter has been patient and constructive through many vicissitudes of writing.

VANNEVAR BUSH

FOREWORD

*This book emerges from a decade of war and uneasy peace;
writing it has been concentrated into the past two years. As I
have been writing, the scene has continually changed, and it
is still changing as the last few words are added. The President's announcement of evidence of an atomic explosion in
the Soviet Union appears as the volume goes to press. The impact of science and the evolution of weapons indeed produce
a fast-moving background for all thought on our destiny. Relative importance shifts, time schedules become altered, international relations evolve into new patterns. Yet the purpose of
this book, and the thesis it presents, are not dependent upon
the quickly changing events of day to day. The principles that
underlie our democratic system will remain the same, the position of science within our framework is unaltered, and the
resourcefulness and steadiness which have carried us through
the tough days of the past decade will carry us through the
future as well, if we preserve and enhance them, whatever the
technical aspects of the problems before us may be. The scene
changes, but the aspirations of men of good will persist. That
there is hope these aspirations may become realities, if we keep
our strength, is the thesis of the book.*

VANNEVAR BUSH

*Washington, D.C.
September 26, 1949*

MODERN ARMS AND FREE MEN

SCIENCE, DEMOCRACY, AND WAR

"There are many sciences with which war is concerned, but war is not such a science itself, and any forecast for the indefinite future presupposes a certitude that is not possible."

—JAMES V. FORRESTAL
Statement before Air Policy Commission. December 3, 1947

ONE QUESTION is in the mind of every American as he faces the blurred future: Will the coming generation of our youth have to fight in another desperate war?

All our opinions, all our acts, are conditioned by this question. We tax ourselves heavily, and more or less cheerfully, in the belief that generous resources will help. We extend aid to Europe, at heavy cost in materials and labor, even at peril to our economy, in order that this aid may help rebuild a bulwark against war. We instituted a peacetime draft, and we strive to unify our military organization, in the hope that if we are fully ready we will not be forced into war.

The central question breaks down into many. Is it true that a new all-out war, with atom bombs and biological warfare, would destroy civilization and drive us back to the dark ages? Is the case so desperate that a prophylactic war is justified in order that we might at least meet the inevitable at our own time and on our own terms? Can a democratic regime develop great military strength without distorting its true nature? Has the time not come when the peoples of the world, in terror before the threat of war, will build one world under law? Or is peace so sweet to those who live at the edge of the abyss as to be bought at the price of chains? What can democracy offer to those in distress toward building a world in which free men will live in concord, peace, and understanding?

1

There are no precise answers to these questions, just as there is no complete answer to the bigger problem of how to avoid another war. There are powerful factors present: science and democracy. Modern science has utterly changed the nature of war and is still changing it. And the democratic process has given us new controls over our destinies that are subtle, only partly understood, and also changing. This book is an examination of the vast process of change in which science and democracy are both affecting the nature of war.

I know no scientific formula with which to explain this great and intricate interaction of old and new forces. Its order of complexity is too high and it has too many variables for either a human or an electronic brain. But it is a process that we can begin to understand, better at least than any generation before ours, and with understanding comes control. For ten years, thanks to the accidents that direct men's lives in a democracy, I was in a position to see as much as any single man could see what science has done and can still do to the art of warfare. It is part of the obligation of any citizen who has been given such responsibility and opportunity as I have, no matter by what accident, to set down for the record what he has learned, and to share with others any light it may throw on the great question of war or peace that haunts us all.

I am not much of a prophet; there is a great deal of guesswork inevitably involved when we attempt to predict just what applied science may still do to our lives. Yet I have specialized in the development of new weapons and wrestled with the scientific problems of total war. I have also watched with fascination the ponderous turning of the wheels of government and the weaving of men's relationships with each other into patterns of incredible complexity. I have evidence that supports the two chief conclusions of this book. I believe, first, that the technological future is far less dreadful and frightening than many of us have been led to believe, and that the hopeful aspects of modern applied science outweigh by a heavy margin its threat to our civilization. I believe, second, that the democratic process is itself an asset with which, if we can find the enthusiasm and the skill to use it and the faith to make it strong, we can build a world in which all men can live in prosperity and peace.

This old challenge has become a new one as a result of the application of science to war in a degree that has completely altered warfare. The combination of science, engineering, industry, and organization during the last decade created a new framework that rendered conventional military practice obsolete. Radar, jet aircraft, guided missiles, atomic bombs, and proximity fuzes appeared while we were fighting; they determined the outcome of battles and campaigns, even though their determining nature was not fully exploited in that contest. Over the horizon now loom radiological and biological warfare, new kinds of ships and planes, an utterly new concept of what might be the result if great nations again flew at each other's throats. It is this which makes the thinking hard.

We cannot take refuge in the assertion that these are matters for specialists: for a State Department to carry on a new form of international discussion, for a Defense Department to prepare us for a new kind of war. We are not a dictatorship, where a single distorted mind names the tune and all the lackeys dance to it. We are a free people, and as we think so will our public servants act.

Must every citizen, then, grasp the full nature of atomic energy, evaluate the modern submarine, predict the consequences of supersonic flight, or grasp the mentality of those who rave at us in councils? It is absurd, of course. But every citizen, in a strange subtle way, visualizes where we are and where he feels we are going, and from this is distilled, in a way we hardly understand, our national policy in every regard. It is not just through votes, or editorials, or commentators that this action occurs, but more indirectly and through a thousand channels.

Since the war ended, this elusive and powerful force, this mass concept, this public opinion, has ruled that we should enter wholeheartedly, in spite of irritations and annoyances, into the attempt to build some sort of United Nations. It has ruled that we should build a strong military machine and have it ready. It has stiffened our backs and frowned on exposure of weakness or overreadiness to compromise. It has most certainly rejected any idea that we should become a conquering nation, or strike early ourselves in the attempt to avert a later and a more desperate war. It insists that we control the traitors in our midst, and somewhat bewilderedly that we do not sacrifice our essential freedoms in the process. It has even begun to insist that special and selfish interests be regulated within a compass that will not wreck the national strength we need. It has placed us on the path we now pursue, yet does not know exactly where it leads.

How has public opinion done all this? In discussion and criticism our people somehow sense out the big issues, of course. But public opinion does this principally because men understand men and, through some process that is still mysterious, select those to be trusted, which is the essence of the democratic process. From the whole seething fracas emerge a national attitude and policy which become the guide for all who manage affairs in the public interest. There is a chain of trust. We do not elect a President because we think he understands atomic energy in all its ramifications; we know and he knows he does not. In November, 1948, we elected a President, in a close contest where two opponents were selected by a system that, illogical as it is, nevertheless produced contestants who truly appealed to us, and we elected one because the majority preferred him. The issues were stated and argued, but did not really determine the outcome. This was determined by the millions and their ballots on the basis of where they preferred to place their destiny. The uppermost thought was whether he who was chosen would lead and judge better, as compared to his defeated opponent, in the interest of the common citizen, marshaling about him those who are specialists, in science, or warmaking, or diplomacy, or politics, marshaling them with a stout heart and common-sense judgment—whether he too could truly judge whom he could wisely trust.

It is highly important that the general outlook of the people be sound as we face the future. If we had been in abject terror, facing a new inevitable war that would destroy our cities, our farms, and our way of life, we would have followed some Pied Piper in the last election who would have led us into the sea. This we emphatically did not do. In spite of alarms, in spite of the prophets of doom, we face the future with resolution. If as a people we had felt all-powerful, that we could speak and the world would tremble, that we had a mission to rule the unenlightened, that we were a superrace, we would have followed a demagogue. There was not even a single demagogue of the sort in sight on the national horizon. The steadiness of purpose of the American people is our hope and refuge.

This national attitude is extracted from a maze of conflicting argument centering about two focuses: the nature of the democratic process and what the applications of science hold in store for us all. The two are intimately intertwined, for science does not operate in a vacuum, but is conditioned by the political system that controls its operations and applications. The discussions on the air or about the stove at the corner store revolve about these two central subjects. They are not always recognized as being present, for the talk may be on the next crop or the ambitions of the local sheriff, but they are in the background. For they determine our destiny, and we well know it. If the democratic process will work to foster an effective government, if science will cure our ills and not merely provide means by which an aggressor can suddenly destroy us, we have a rosy outlook and can quarrel about minor things without fear. If we are headed over a cliff, with our means of progress out of control; if our form of government is transitory and bound to transform itself into selfish rule by some dominant group; if the application of science has finally doomed us all to die in a holocaust, there is little use in arguing about the drought or the next strike. The two central matters are interconnected. What science produces, in the way of applications within its own changing limitations, depends upon what is desired by authority, by those who rule or represent a people. Pure science may go its own way, if it is allowed to do so, exploring the unknown with no thought other than to expand the boundaries of fundamental knowledge. But applied science, the intricate process by which new knowledge becomes utilized by the forces of engineering and industry, pursues the path pointed out to it by authority. In a free country, in a democracy, this is the path that public opinion wishes to have pursued, whether it lead to new cures for man's ills, or new sources of a raised standard of living, or new ways of waging war. In a dictatorship the path is the one that is dictated, whether the dictator be an individual or a self-perpetuating group.

When for the first time in history the decision was taken to recognize scientists as more than mere consultants to fighting men, I was living in Washington, and President Roosevelt called on me to head the job. Abraham Lincoln had set up the National Academy of Sciences during the Civil War, and Woodrow Wilson had authorized the National Research Council during the First World War. Both did good work, and their histories are illustrious, but neither was given large funds or authority. In the National Defense Research Committee and the Office of Scientific Research and Development, in the Second World War, scientists became full and responsible partners for the first time in the conduct of war.

We had, during the war, approximately thirty thousand men engaged in the innumerable teams of scientists and engineers who were working on new weapons and new medicine. We gathered the best team of hard-working and devoted men ever brought together, in my opinion, for such a task. We spent half a billion dollars. Congress gave us appropriations in lump sums and trusted us to decide on what projects to spend the money. In uniform but without insignia, some of our men were on the battlefields and in the planes and ships in every theater of war.

At the same time, I watched a great democracy bend itself around this new development and give it life and meaning. In this I was a rank amateur, knowing little about the details of the democratic process but believing in it. I saw it work. We contested with generals and admirals, but the new weapons were produced and used, and we wound up friends. We argued among ourselves, but of roughly thirty-five men in the senior positions in the scientific war effort only one man was absent when the war ended, and this was because of illness. We did a job that required fantastic secrecy, and yet we won and held the confidence and support of the military, the Congress, and the American people. We were a varied group with all sorts of backgrounds and prejudices, and yet we developed a team technique for pooling knowledge that worked.

This is not a history of what science did in the war; that has already been written. It is an attempt to explore its meaning in the relations between man and man, as individuals and in the organizations they create. Since the beginning of organizations there have been two controlling motivations that have held them together. One is fear, utilized in the elaboration of systems of discipline and taboos. The other is the confidence of one man in another, confidence in his integrity, confidence that he is governed by a moral code transcending expediency. Most governing organizations have involved a mixture of these motivations; they always will as long as the nature of man remains unaltered, but one may be controlling and the other subsidiary, incidental, or extraneous. There has been a general feeling that the second is the higher motivation, but that it is inherently weaker in dealing with the harsh and complex conditions of existence.

The subject is of extreme seriousness to us today. The world is split into two parts that confront each other across a gulf. On one side is a rigid totalitarian regime, ruled through fear by a tight oligarchy, which sees only two possibilities: it will conquer the world or succumb in a final struggle. On the other side is a diverse group of nations, with democracy as the central theme, which aspires to a world of peace under law and would bring it about by advancing collaboration and mutual confidence, respect for the given word, integrity, as higher and essential bases of action if the world is to be more than a mere police state.

Neither is absolute in its nature. Within the totalitarian regime there is still an aspiration for freedom; there is, moreover, in the great mass of those rigidly controlled from above an idealism, a neighborly helpfulness, a grasp of something higher than selfish ambition, which still persists in spite of regimentation, propaganda, and the evils of the secret police. Within the democracies there is still plenty of chicanery, a negation of principles in the treatment of minorities, abuse of the necessary police power. Yet the issue is still clear. On the one hand is an absolute state, holding its people in subjection and molding them for conquest by force or trickery. On the other, there is hope of better things.

Through the pattern of modern thinking runs a doubt, a question as to whether a system based on the dignity of man, built on good will, can be sufficiently strong to prevail. The thesis of this book is that such a system is far stronger, in dealing with the intricate maze of affairs that the applications of science have so greatly elaborated, than any dictatorship. The democratic system, in which the state is truly responsive to the will of the people, in which freedom and individuality are preserved, will prevail, in the long run, for it is not only the best system, the most worthy of allegiance that the mind of man has built; it is the strongest system in a harsh contest.

The striking success of the application of science in controlling nature for our purposes has not only modified the conditions of the contest, producing radio for propaganda in a cold war, and atom bombs to consider in estimating the consequences of allowing it to become hot; it has also modified the underlying philosophy with which men approach the problems of their organization and government and every other aspect of their existence. Totalitarianism and tyranny, and the struggle of men for freedom, existed long before science became widely applied. But the success of science has given concrete form to the clash of philosophies that now divides the world. On the one hand, fear is seized upon as the only dependable motivation, and moral codes are discarded for a blatant expediency, because of a crass materialism that has become embedded in a totalitarian regime, and this regime enthrones science as its model. On the other hand, there is faith, even though at times it may not extend beyond faith in the dignity of man.

The philosophy that men live by determines the form in which their governments will be molded. Upon the form of their government depends their progress in utilizing the applications of science to raise their standards of living and in building their strength for possible war. Upon this form depends the effectiveness with which they can provide for security against the ravages of nature or of man. Upon the form depends also their progress in securing justice and maintaining opportunity. Upon it depends the outcome as to whether life is worth living.

Through it all runs the thread of the impact of science in altering the world and the relations of men therein. It will not be enough to trace the current and future development of weapons or even the ways in which science may further alter our material affairs. We need to delve deeper, and we shall.

CHAPTER II

THE TECHNIQUES OF WORLD WAR I

"This helplessness of the art of war, we can see now, was due to the slowness with which armies changed away from an old conception of warfare to a new one. All the technical means for ending this helplessness were present early in the war; the gasoline engine, the caterpillar tractor, the idea of an armored vehicle capable of crossing trenches and standing machine-gun fire, the aeroplane and the light machine gun were all available. What was not available was the idea of war as a changing art of science affected by every change in the techniques of production and transport, and inevitably out of date if these techniques were not employed to the full."

—Tom Wintringham
The Story of Weapons and Tactics. 1943

A NEW ERA in warfare started with the First World War. Two great innovations were responsible for this, each of them destined to have results as far-reaching as the first bow, the Romans' siege engines, the introduction of gunpowder or of the rifled gun and the shell. The first of these new departures was the development of precision manufacture and mass production. The second was the internal-combustion engine. Between them, they made mechanized war possible, and the world will never be the same again.

They changed the nature of war. Their applications covered the entire range of weapons and the whole strategy of war. Consider for a moment two of them: the machine gun and the submarine. Each of these was something new; each of them profoundly changed the way in which nations carried on the ancient business of fighting.

On land, the First World War became a deadlock very early. It came as a surprise to most of those who had learned their strategy and tactics by studying the battles of the past. But entrenched men armed with machine guns and shielded by barbed wire could not readily be dislodged by any means in existence, and battle lines became bogged down in the mud for years. With the horrible features of a war just ended fresh in our thoughts, we find it difficult to recall those lines in France, but of all the hardships and nerve-racking stresses that are part of all war there probably was never a more distressing phase than that of millions of men locked in combat in the mud of Flanders. For our purposes we note one primary point: the creation of complex automatic mechanical devices in quantity, the machine gun and barbed wire in particular, ended forever the hot rush of masses of men and replaced it with doggedness, a new kind of courage and endurance, a skill at operating machines under stress, and for the first time the factory behind the lines became a dominant element in the whole paraphernalia of war.

The means were available to break the deadlock, but they came late and on an utterly inadequate basis. The same mechanical processes that produced wire, artillery, or the machine gun could also build tanks, and the internal-combustion engine using petroleum products was available to propel them. The tank could roll down the wire field, and men could follow. But military men, and in the democracies the civilians who held ultimate authority, were with a few notable exceptions far from alert to the trends of the times, and the tank became an orphan that received scant support, so that it appeared in quantity only when exhaustion had already made the deadlock tenuous.

The internal-combustion engine—that convenient device which packaged great power in small space and weight—had also made the aircraft possible, and men had flown for a decade before the war began. Yet the participation of military aircraft in that war was an incident rather than a determining factor. If there had been vision and an application of the true state of the art, they too could have broken the deadlock, for they were almost immune in the absence of antiaircraft artillery, but they served principally for reconnaissance.

So land warfare bogged down, and men said that now the defensive was permanently in advance, and that all future wars would be of the nature of sieges, with control of raw materials the prime determinant. It was true at the moment, but the seeds of a vast change were present.

There was another innovation that appeared in land warfare and that we need to note because it was the forerunner of new things—poison gas. It appeared because mass chemical manufacture had become possible and because the close-locked nature of the combat gave it opportunity. There are two aspects of any weapon, old or new, for destroying the enemy or his facilities: the destructive means itself and the means for its delivery or dissemination. Poisons, powerful poisons for that matter, had been available for many years; the Medicis, among others, in Italy in the fifteenth century had been adept at their preparation. Poison gases themselves had been known for a long time; one of them, chlorine, was readily prepared and used to purify water and could be made in quantity at relatively low cost. Chemical ingenuity soon produced new ones, notably mustard gas and lewisite, and these were potent indeed. But the real factor that brought them into use was the close proximity of great bodies of men. Gases were lobbed over the lines by simple mortars, or merely allowed to drift downwind. It is notable that they were not to any appreciable extent spread by aircraft, even though aircraft could fly over the lines almost at will, slowly and at low altitude, as though dusting crops. Toxic gases produced results, horrible ones at times, and added to the hazards and discomforts of positional warfare, but they were not determining, and they did not break the deadlock. The deadlock finally broke as the German armies were defeated in the field and cracks appeared behind the lines in the civil organization, and the war on land ended with important and far-reaching new instrumentalities in existence, but with scant appreciation of the fact that war had become a new thing that involved all the efforts of a nation and in which industrial potential and the resourcefulness of scientists and engineers were primary factors in determining the outcome.

On the sea also there was a beginning of a revolution in methods, but only a beginning. Great fleets still met and slugged it out with guns, and their power depended primarily upon intricate devices for aiming those guns, although even these were not fully exploited and there was still dependence upon rugged seamanship and the sighter's eye. In fact, the introduction of fire control into our own Navy had been accomplished only after vigorous action. Actually, successful warfare at sea had become a matter of having better mechanical and electrical devices than the enemy, with men better trained in using them, so that they would function accurately under the stress of battle. Potent seeds of progress were present here also. The computing device that aimed the great guns of the battleship automatically allowed for the course and speed of ship and target; it introduced into its computations the deflection of the shell owing to the rotation of the earth, the effects of gravity, and the barometric pressure. It utilized the gyroscopic compass, that delicate device which senses the rotation of the earth and thus points out north with far more positiveness and accuracy than the old magnetic compass. It aimed guns with much greater precision than could be accomplished by any number of skilled men alone.

There were here the beginnings of two new and powerful concepts. First was the idea of reliable complexity in intricate devices, in masses of electrical and mechanical parts interconnected to function in precisely predetermined manner, and dependable in spite of their intricacy because of the contribution of standardized mass-production methods. The second idea was that of relegating to a machine functions of computation and judgment formerly performed by men, because the machine could work more rapidly, more accurately, and more surely under stress. These ideas were bound to be applied to many tasks, especially to war, where speed and precision in complex tasks are foremost. They would in time have produced a new

sort of battleship, or floating fort, except that the days of the
battleship were coming to an end. They ultimately produced a
host of devices for mechanized warfare of all sorts.

The great innovation of sea warfare was the effective submarine. A submarine had sunk a ship during the War Between
the States in the 1860's, but the undersea craft reached effective
form only through the advent of the internal-combustion engine
and the storage battery. It was thoroughly underestimated when
the First World War broke out; there were in existence practically no means of combating it, and, despite its crude form, it
nearly determined the outcome of the struggle before it was
finally overcome by the depth charge and the convoy system.
It came near to cutting the world apart; if it had done so the
Kaiser's dreams of conquest would have come true. The submarine was then slow, could not submerge deeply, and was
fragile, but it had a deadly weapon, the torpedo, and this too
was destined to lead to important advances. The torpedo had
long been known; it was itself a submarine, propelled by a steam
engine, automatically controlled to run at a fixed depth, and
equipped with a gyroscope to cause it to run in a straight line.
It carried a charge that was bound to be lethal to any merchant
ship and highly damaging to any warship. The cargo ships roll-
ing down the seas, so necessary if Europe and England were to
survive, were sitting ducks for these undersea raiders until con-
voying was introduced, and even then their survival was touch
and go. The submarine was barely overcome in that war, though
it was then in only a crude state of development, with great
opportunity for improvement.

There was an exceedingly important lesson here. Mahan had
taught, and taught effectively, the importance of control of the
seas. His teachings were sound, but they depended upon two
premises that time and technical innovations had changed by
the time of the First World War. In his arguments he postulated
nations dependent upon sea commerce for their safety, their
prosperity, and even their continued existence as powers, and
postulated that control of the seas was dependent upon the clash

of surface fleets. From these arguments it is readily deduced that the mission of a fleet is first to destroy the opposing fleet and then to cut the enemy's lines of supply, thus throttling him. But conditions are altered so that the postulates no longer apply in simple form when there are fleets that can destroy but cannot meet in battle—the characteristic of submarine fleets. They are also vitally altered when there are land powers which are internally self-sufficient, or nearly so, by reason of synthetic materials and the like, but which, having access to the sea, can maintain submarine fleets. For, if there is no answer to the submarine, such a power can cut the rest of the world apart at will, in spite of great fleets of battleships or other craft, while it itself remains immune, and can thus impose its will upon the world. Germany under the Kaiser nearly did so.

This lesson was fully apparent when the First World War ended, for it was obvious that the somewhat crude submarine and torpedoes of the day could be greatly improved. Improved, they would have been formidable weapons for the future, with the advent of air power and particularly the marriage of sea and air power. Yet the lesson remained and was nearly forgotten in the years that followed.

Radio also appeared in this first war. The embryo of great developments, it was useful primarily for communication at sea and to correlate movements in sea battles, but it presented in use most of the beginnings of what led later to the extraordinary ramifications of what we now call electronics. This is that branch of the science and technology of electrical phenomena which uses equipment depending on the controlled flight of electrons and ions in space. Radio tubes and television tubes are examples of electronic equipment, and electronics embraces all that we have so far learned about control of the flight of electrical charges.

One of the salient elements in this was the thermionic tube, probably the most versatile device in an entire generation of rapid electrical progress. It is, essentially, a vacuum tube in which the flow of electrons is controlled for purposes such as detection or amplification. Thermionic tubes perform functions in ordinary radio sets, but they can accomplish a multitude of other things. The thermionic tube was joined by the cathode-ray tube, in which the electric flow is a beam of ions from a cathode, impinging on a fluorescent screen to make a picture or to present visually the performance of other apparatus connected with it. This is the foundation of television, as it is of radar. Many other devices followed, to increase power and versatility: thyratons, which are like thermionic tubes except that they use ions in a rarefied gas instead of electrons in a vacuum; photoelectric cells that can substitute for the human eye; relays of many forms for ordering the sequence of events in electrical circuits. These circuits could now be constructed to perform intricate computations or sequences of reactions. They could be combined with mechanical masses of gears and levers. They could respond to light or sound or radio waves, and they could move levers, steer vehicles, or fire explosives. The important point was that such combinations could do things a man could not do, they could go where a man could not be sent, they did not become confused in the stress of battle, and as techniques improved they became reliable and relatively inexpensive when produced on production lines.

When the First World War ended there were thus in existence nearly all the elements for scientific warfare. The principal devices had been tried out in practice. There were automatic guns, self-propelled vehicles, tanks, aircraft, submarines, radio communication, poison gases. More important, mass production had appeared; complex devices had been made reliable; the petroleum, automobile, chemical, and communication industries had approached maturity; thousands of men had become skilled in techniques. The long process of applying scientific results, all the way from the original academic theory or experiment to the finished device, had become ordered. The world was fully launched on mechanized warfare. For all the technical devices that were later to be used in the second war, except only atomic energy, practically every basic technique had appeared, waiting only construction and development. And this was in 1918.

CHAPTER III

BETWEEN THE WARS

"During the two decades between the end of World War I and the beginning of World War II, the people of the United States had pinned their faith on the impossibility of another world war, although their government had not been willing to bear any share of the responsibility for the success of the one international organization which had the slightest chance of making aggression unlikely." —IRVIN STEWART
Organizing Scientific Research for War. 1948

WHAT DID THE WORLD do about it? It went to sleep on the subject. In this country, a decisive factor was the general atmosphere of isolation; here and elsewhere in the world there was a feeling—closer to hope than to conviction, but still a powerful feeling—that great wars were over. Fundamentally, lethargy gripped the technique of warfare between the First and Second World Wars. Those who were familiar with modern scientific trends did not think of war, while those who were thinking of war did not understand the trends.

So the Second World War began where the first one ended. There were a few exceptions. A very few new technical weapons were worked on, which will be mentioned later. There were a few small wars, like the one in Spain, which were regarded by some military men as practice in the techniques of modern total war. Development of new machines for commercial purposes, of course, produced astonishing results, which could later be exploited for military purposes. But despite the change in the whole nature of war that was obvious in 1918, there had been almost no serious exploitation of its technical lessons by 1939.

There are plenty of examples of how all the nations failed to apply modern scientific techniques in their preparations for war. The best is that of Hitler. When he first came to power in 1933,

17

he proceeded to destroy the great structure of German science. He did so by eliminating those scientists who did not fit into his distorted racial or political concepts and by regimenting the remainder. The fundamental scientist can do little of a practical nature alone, but he is an essential link in a chain, and this fact Hitler did not understand. It is fortunate for the world that dictators are very likely to be obtuse, and beyond influence or conversion, when it comes to the subtle ways in which science, engineering, and industry are interlinked to produce more than obvious progress in any field, and especially in the art of war.

During the first war no country had an effective organization for joint functioning of scientific and technical men on the one hand and military men on the other. Scientific and technical men did not sit on war-planning councils, and military men in general regarded scientists and engineers with either forbearance or contempt.

There were some bright spots, of course, in the First World War. In this country a group of scientists working through the National Research Council developed, in 1917, improved listening devices for antisubmarine warfare by which the elusive submarine could be better detected and followed, and this helped somewhat. There were groups in England and France similarly engaged. It is fortunate that there was plenty of doggedness and rugged seamanship in this first contest with the submarine, with tools that remained crude in spite of devoted but weakly supported scientific effort. Otherwise England would have been starved and there would have been no American Expeditionary Force. Even in this critical aspect of combat, the method of destroying submarines after they had been detected proceeded no further than dropping a can of explosives overboard, and the methods of detection did not proceed far.

Largely, however, the First World War was fought with the weapons that existed at its outset. There were plenty of military men who held that all wars had been and would be thus fought; and some of these were still in positions of high authority as the second war got under way. In spite of the presence of the elements for mechanization of war, the first war was little mechanized, and when it ended progress along those lines nearly ceased.

Certainly this country failed to make much progress in the application of science to military matters. Congress cut appropriations, while the Army and Navy applied what they received principally to other things than improving their methods. Industry was not interested and took military contracts with reluctance. The reluctance was justified, for the business was unattractive, and the head of any company that developed weapons was likely to be called a merchant of death. Science turned its full attention elsewhere. Technical men in industry were developing some of the most bizarre gadgetry the world had seen, but not for war. In this country it was not merely that the people turned aside from the paraphernalia for war. Civilians felt that this was a subject for attention only by military men; and military men decidedly thought so, too. Military laboratories were dominated by officers who made it utterly clear that scientists or engineers employed in these laboratories were of a lower caste of society. When contracts were issued, the conditions and objectives were rigidly controlled by officers whose understanding of science was rudimentary, to say the least. To them, an engineer was primarily a salesman, and he was treated accordingly. Undoubtedly these blocks to progress existed in other parts of the world as well as in this country. At any rate, almost no progress on military devices emanated solely from military laboratories or military men.

Here and there, intelligent and aggressive groups of military men or civilians made progress in spite of these obstacles and the general atmosphere of condemnation of thought of war. Some of these need to be noted.

Science pursuing its own ways, and industry bent on making a profit, produced advances between the wars that were to influence methods of warfare profoundly. In general these advances employed elements that existed when the period began, but there were significant additions.

The rise of the automotive and radio industries was striking. It need not be reviewed in detail, for it is familiar to us all. The direct results, in the creation of better automobiles and trucks, better radio sets and the beginnings of television, were important in themselves, but these were exceeded in ultimate influence upon war by the indirect results. Manufacturing methods took great strides, first along the lines of interchangeable precision manufacture of mechanical apparatus, and second in assembly-line methods for producing intricate electrical or mechanical gadgetry in quantity. Still more far-reaching was the effect on the attitude and capabilities of the people. The fact that there were literally millions of youngsters who could drive cars, or repair them, who could build their own radio sets and communicate as "hams" all over the world, a whole generation of competent resourceful mechanics and electricians, was the best insurance that could have been produced for the strength and safety of the nation in a world of modern war. Every corner garage, every radio club, was a sort of center of training, training that could be readily transformed in a short time, when the test came, into ability to operate the complex implements of war. Any explicit training in camps, under military supervision, would need to be very intelligent indeed to beat it. ·

During the second war, for example, one radio "ham," whose formal education had been limited to grammar school, helped fight from a laboratory. He was a mechanic; before that he had worked in a spool mill; his father had been killed in a sawmill accident when the boy was four years old. He had picked up his knowledge of radio while he made his living, just as millions of other American boys still do. He became the principal designer in this country of magnetrons. A magnetron is a type of thermionic tube in which part of the control is magnetic, and it is the very heart of radar. He can talk today with Nobel Prize physicists, and can understand them and tell them things they want to know.

The greatest peacetime development, from a military standpoint, was in the field of aeronautics. Here there were many forces at work. Primary was the coming of age of air transport, with plenty of vicissitudes, without much aggregate profit, but in the hands of true pioneers who did not shrink from taking a risk and did not settle into their collars when a technical innovation was proposed. Second, there was genuine military interest here, and real stimulus by military orders and development, fortunately on a novel basis, whereby military men stated requirements in general terms but left private industry free to meet them with its own flexibility and resourcefulness. This was a new thing under the sun, and a potent one. It occurred because those military officers who pioneered in aeronautics in the first war were young and unconventional and remained so after the war ended. In fact, they are still young and unconventional, even today, and while this fact now has its disadvantages, in the early days it was essential to real progress. Those military aeronautical pioneers were essentially rule-of-thumb or handbook engineers, men who came up through the roundhouse, to write in railroad terms. No disparagement is intended; they were just the sorts of individuals needed as the art then stood. They would try anything, at least once. Their organization was loose, and, as was bound to be the case in view of their individual characteristics, they avoided the rigidity that often keeps military affairs in a strait jacket. We owe them much.

If they had been the only actors in the scene, however, aeronautics might have gone down the road that was taken by many an industry that depended upon empiricism and failed to understand research, where men grew old and stodgy yet still retained control in the belief that all real progress was made when they were young. This condition would have been fatal, for the science of aeronautics was just beginning, and the art of flying, more perhaps than any other, needed to be placed on a sound scientific basis. Fortunately, however, there were other protagonists. One of these, already mentioned, was the engineering and design groups in private competing companies, who were

kept alive by the pressure of competition and were highly realistic under the necessity of making a dollar in a tough industry.
The other was the aeronautical scientist.

Aeronautical science moved ahead well, thanks to a strange
set of circumstances. There was need for fundamental work in
aerodynamics, but there was also need for much applied research, linking theory with design, and requiring for its furtherance great and expensive wind tunnels, beyond the resources of
either the universities or the precarious industry.

This situation might well have resulted in a great military
laboratory, under the control of officers in rotation, with a civilian
staff. Now such laboratories have their place; there are certain
things that can be accomplished by no other means, and the
skill of managing them effectively is gradually being acquired
by the military departments. At the end of the first war, however, they were deadly establishments. That the development
of aeronautical science did not fall primarily into this pattern
was owing to the creation of the National Advisory Committee
for Aeronautics. It was formed during the first war, on a decidedly small scale, largely because of the vision of the grand
old man of aeronautics, Dr. William F. Durand, who was
largely responsible for the initiation of the whole aeronautical-
research program of this nation.

It slowly grew into a great enterprise, with mammoth laboratories and a diversified staff. The extraordinary aspect of the
NACA is that it is a part of governmental organization, independent of the military, reporting to the President and Congress, and governed by a body of independent citizens appointed
by the President and serving without compensation. Military
representatives, and representatives of government bureaus
outside the military departments, sit on the committee, but
the control of policy has always been largely determined by the
citizen members. Its research has for a generation laid the
groundwork for aeronautical advance in this country. Today
the universities have joined in extensively, and there is a pattern of interrelation and support that operates to advantage.

But in the pioneering days there would have been little scientific basis for advance if the NACA had not existed.

The record of the advance in aeronautics, from the old biplane held together by wires to the sleek modern ship of commerce, from plodding speeds to supersonic flight, from low altitudes into stratosphere, from a hazardous sport to extraordinary and mounting safety, from erratic performance to reliable transportation, is an intricate story of the application of science to the art of flight.

Not all the advance occurred within aeronautical circles, of course. A modern aircraft draws upon many fields of effort. The full advance could not have occurred, for example, if communication industries and techniques had not been advancing simultaneously, to provide navigation means, control systems, and the like. Hydraulic systems for landing gears, tires that would stand the racket, automatic pilots, a hundred elements, appeared because techniques could flow between one field and another. The result was good commercial aircraft and military aircraft of the advanced nature with which we entered the second war.

The engine is the heart of the aircraft. Successful aircraft appeared only when advances in other lines made it possible to pack large power into restricted space and keep weight down. The development of the engine paralleled that of the aircraft itself. Here the influence of the NACA was not great, nor was university research directly on engines highly important. The advance was made by the engine manufacturers themselves, on the basis of commercial and military requirements and orders. It was profoundly affected by improvements in metals and fuels, the former because of modern metallurgy, and the latter because of the growth of the petroleum industry and its research programs. It was also affected by the automobile industry, but the building of really fine engines went far beyond the requirements of this industry. The automobile does not actually need an engine of really advanced refinement and cannot afford it, so automobile engines are somewhat elementary and prosaic.

For aircraft, however, engines became marvels of precision and endurance. When we consider the inherent complexity of a reciprocating engine, with its myriad of precisely fitting parts, its high temperature and stresses, its fuel mixers and ignition, and the power it packs, and consider also the confidence with which we have come to depend upon its continued reliable performance, it becomes one of the greatest exemplifications of man's skill in mechanical fields.

. But, even so, we went down a conventional path in engine development between the wars, merely refining the old basic designs, and jet propulsion and turbo-jets did not appear until the stress of new war brought them forward. There is some advantage to inquiring why this was so. The facts are that even large private enterprise has its limitations on innovations when they are off the beaten track and expensive, and that government programs have in general been much more seriously limited in this regard. When there are small competing industries, when new ground can be broken without too large financial risk, sometimes when there are individuals with the itch to take a chance and with the rare combination of that initiative with common sense, then sudden departures from the conventional will occur, and whole new instrumentalities or industries will open up. It is not simply because someone suddenly invents—people are always inventing. Though a smoothly operating patent system is essential if we expect private funds to take long shots for our benefit, the opportunity to patent an invention is not enough. There must be initiative, brains, and willingness to run a risk. Really striking movements away from the crowd will not occur once in a thousand years in a tight bureaucracy. We can take comfort in the conviction that dictatorships will seldom pioneer, and that when they do the dictator will probably buy gold bricks; but our subject at the moment is the technical advance in the art of war during the interval of peace, and the ways of organized men with respect to untrammeled thinking and action will have to wait. Between the wars this country developed magnificent aeronautical engines, but it did not put the gas turbine in the air, even though the basic idea was old, the materials and necessary theory were largely available, and the opportunity was great. Military officers, industrialists, engineers, members of NACA, all missed a bet. There is cold comfort in the fact that all other countries, totalitarian and democratic, missed it also.

During the peace the techniques of artillery moved ahead in the slow, plodding way in which the art of firing projectiles has always moved since the days when guns were first rifled and explosives were first placed in shells. There is something about the word ordnance that produces stodginess in its adherents. Again this is not a matter of this country alone; it seems to be a general affliction. Gun-control systems improved somewhat, but this advance was hardly remarkable in view of the fact that extraordinarily successful and complex computing and control systems had been in use on naval guns in the first war. These were applied to antiaircraft artillery, but this change was not a severe stretch, and industry under government orders made the transition largely by adapting previous practice. Even this advance did not occur everywhere by any means, and in this field at least this country was well ahead. Very haltingly, guns were placed on caterpillar treads to render them mobile, although the earth-moving industry had long been moving all sorts of things this way. Guns were made bigger, of course, and muzzle velocities and precision increased somewhat. But the whole gamut of new ordnance devices—rockets, recoilless guns, guided missiles, proximity fuzes, bazookas, frangible bullets— waited for the pressure of war, appearing then largely outside the organized system of ordnance development, and sometimes in spite of it.

During the interval of peace two military devices of importance were developed in governmental laboratories and associated industry and kept very secret. They came about because there were knots of able and vigorous officers who forged ahead, kept in touch with civilian progress, saw uses for military purposes, and pushed them to test and acceptance in spite of solid brass and hazards to careers. They were radar and sonar. We shall deal briefly with them later.

Yet, even though military development was spotty we were during the peace becoming prepared for war, not consciously or deliberately, but as a result of building a new industrial country. With it went a considerable advance in American science, fundamental and applied. We had always leaned on Europe for our basic science, devoting our attention rather to applications and gadgetry, but now sound extensive basic research programs began to appear. These were centered in our universities and research institutions, private and state-supported. In the same interval one industry after another began to move out of cut-and-dried empiricism, or plain somnambulance, into deliberately planned programs, using the science applicable to its field. Engineering education expanded and became more vigorous in its methods. Larger numbers of our population became educated, in school or in the hard pattern of experience; we were an able, versatile people, competent to deal with modern devices of any sort. We were prepared for war—if it did not come too suddenly.

And all the time, in the interval between wars, the science of the atom moved steadily forward.

CHAPTER IV

THE TECHNICAL WAR
ON LAND

"American armed strength is only as strong as the combat
capabilities of its weakest service. Overemphasis on one or
the other will obscure our compelling need—not for air-power,
sea-power, or land-power—but for American military power
commensurate to our tasks in the world."

—GENERAL OMAR N. BRADLEY

THE SECOND WORLD WAR was, far more than the first, a war of
applied science. The great campaigns that swept across the
continents and the oceans drew so heavily on the accumulated
stockpile of fundamental scientific knowledge that this was all
but exhausted when fighting stopped. Nearly every nook and
cranny were explored in the application of theoretical scientific
knowledge to the weapons of war. Should a new war have to
be fought in a decade or so, there will be innovations, but in all
serious probability no such burst of new devices as accrued
when organized science and engineering first turned their full
effort into war, drawing without inhibition or restraint upon the
great unused accumulations of the past. For a new war farther
in the future, the probability of major innovations is propor-
tionately greater.

The first main conclusion we can draw is that there is no
longer any such thing as exclusively land warfare. Amphibious
operations and the interaction of air operations with the movement of armies ended all that. But we can nevertheless look at
land campaigns as a phase of warfare.

When the war first opened actively, after Hitler had consolidated his conquests in the east while the Allies waited behind

the Maginot Line, it soon became evident that the deadlock on land, characteristic of the first great war, was virtually a thing of the past. Masses of mobile artillery could prepare holes, and cavalry in the form of the tank-air team could exploit them. Hitler conquered the Continent because his army had so perfected this method that it overwhelmed customary resistance. The air effort took the form of dive-bombing to remove nests of antitank guns impeding the advance. An exceedingly important feature was the fact that radio communication between ground and air really worked, and the organization was such that a blocked ground unit could call for air support without going through a long chain of command. One may contrast this with the time when the British were operating much later off Norway and field units could communicate with each other only by way of London. The rush over the Continent would also have engulfed England, which was almost defenseless, except that amphibious methods had not then been developed, the British Navy was still in existence, and British spirit and determination were magnificent. The rest of the war on land, from a technical standpoint, focused about these two aspects: attempts to restore or further to break the defensive line, and the elaboration of amphibious methods.

After a long interval, the war on land ended with the same sort of sweep over Europe with which it had begun—the breakthrough of lines, the wide cavalry sweeps, mobile warfare in all its rush and complexity. From this fact it would be easy to jump to the false conclusion that there had been no real change in the interim, that the deadlock on land of the first war had been permanently and universally broken in the early part of the second war. Actually, the facilities of defense had improved enormously, and the means of successful break-through in 1940 had become utterly obsolete by 1944. In the final break-through many new factors were present: long preparation by bombardment of communications, exhaustion of the enemy after five years of war, the finest armies the world has seen, effective coordination of allies in the field for the first time in history, unprecedented industrial production, and magnificent leadership.

The question of how matters now stand, in regard to a deadlock on land, may be argued at length. It is certain that the fully prepared lines of a competent industrialized nation are not going to be broken by an enemy of equal size unless the latter is capable of operating effectively in the field large masses of highly advanced technical equipment, and this is true even assuming the presence of atomic bombs in moderate quantity on both sides.

One of the new primary aids to the defense that appeared in the interval was the use of great quantities of land mines. A field of such mines, covered by antitank guns, distributed in depth, interspersed with machine-gun nests, backed by artillery, is a formidable defense line indeed. In the contest between land mines and means for removing them, the mine won out. Portable devices for detecting metallic mines were successfully developed; they worked along much the same lines as the defective radio set that whistles when one waves his hand near it, but they were avoided by the plastic mine. Dogs were taught to smell out mines, and they did, but there are too many ways of tricking dogs for this to be of much use. Great rollers, pushed ahead of tanks, capable of withstanding mine explosions, had some success, but not much; an occasional very large mine could wreck them. Snake—pieces of hose full of explosive, capable of being pushed ahead or pulled ahead by a small rocket—could be exploded to clear a lane. Tanks equipped with a succession of these devices could and did proceed a way, provided the tank itself was not destroyed, but this was a laborious method indeed for fields of great depth. Very light vehicles, in particular the Weasel, a treaded vehicle of low-unit pressure originally developed for use in snow fields, could proceed over mine fields set for heavy vehicles, but this machine was not armed or armored, and mines set for light pressure could stop it. The Russians apparently merely ignored the fields, moved ahead, and accepted the losses. These could be very large, for example when mines were built to project a can of explosive into the air to explode there and spread fragments over an area. The ratio of losses on the eastern front as the Russians advanced reflects this fact. The days when hordes by their mere number could overwhelm fully prepared positions approached their end.

Moreover, a defensive element entered, toward the end of the war, that the Germans did not have and that added enormously to defensive possibilities. This was the proximity fuze, which we shall discuss in other connections, but which is considered here as used by artillery against ground troops in the open. The old method was by timed fire, that is, by a fuze in the shell timed to cause it to explode at a chosen distance above the ground, whereupon the shell would spread its shrapnel in a deadly cone directed nearly downward. Timed fire could be used only when one could see the ground, and it was not very precise. The proximity fuze contains a little radio set that triggers off the shell at a predetermined distance above the ground. No observation of bursts and no difficult presetting are necessary.

The advent of this fuze increased the effectiveness of artillery against personnel in the open by a large factor, perhaps as much as ten times, and this gain was equivalent to having ten times as much artillery at work. When artillery effect is multiplied against moving men, and not for its effect in blasting an enemy's fixed defensive installations, there is a strong turn in favor of the defense. The proximity fuze appeared in land use just at the beginning of the German counteroffensive in December, 1944, at the Battle of the Bulge. Especially on the northern side of the German break-through it caught German troops in the open, in the fog, at road intersections, as they advanced, and it spread consternation. Its power was by no means fully exploited; this was a first use, and artillery officers, and especially the high command, by no means appreciated what had been placed in their hands. Yet, by means of this fuze, together with air effort as the fog lifted, the lack of fuel on the German side, determined resistance and defiance exemplified best by the historic remark of the commander at Bastogne, and adroit movement of divisions, the break-through was stopped before it could be effectively exploited. The proximity fuze may well have saved Liége.

A mine field, adequately covered by artillery or mortars with proximity fuzes, is a formidable line. Mere combinations of tanks and men will not penetrate it.

The tank, moreover, nearly met its match in the last war; perhaps it should have. A limit to the practicable size of tanks and the thickness of armor that they can carry is imposed by the necessity that they shall not bog down in the softest going to be encountered. An explosive charge can be constructed, and a relatively small one at that, that can penetrate the thickest armor they can carry. This is because of a very old principle called the Munroe effect. A mass of explosive, properly shaped to focus its effect, triggered off at the right instant and in the right manner, will bore a most extraordinary hole in steel or in anything else. Through the hole goes a deadly blast. This principle was used in the bazooka, a light rocket-firing weapon that could be carried and fired by one man. Its muzzle velocity was very low, its precision was poor, it was dangerous to use, but it could stop a tank, and its advent was new strength for the infantry.

Another innovation that spelled difficulty for the tank was the recoilless gun. This was a gun that shot forward and backward at the same time. Such a gun sounds like an inventor's lurid dream, and there were many such, but the forward effort could be caused to give a projectile good muzzle velocity and precision in a rifled barrel, while a backward blast through an orifice removed the recoil and was harmless unless someone inadvertently got close behind the gun. The importance of this innovation resided in the fact that it produced a powerful, relatively inexpensive weapon, of low silhouette, readily moved and used by a couple of men, and hard hitting even at a moving target at some distance. This weapon and the shaped charge were never combined. There seems to be no inherent reason why

they should not be. A tank wandering through a country in-
fested with such weapons would have a short life.

Another formidable enemy of the tank is the rocket-carrying
plane, although this weapon works both ways and can be used
to support as well as oppose tanks. The relatively slow dive-
bomber with which the Germans first swept forward soon became obsolete. The high-speed plane, necessary for reasonable
avoidance of flak over protected lines, with relatively flat flight,
could not place a bomb precisely on a small target such as a tank
or a pillbox. But the rocket-carrying plane could. The velocity
of the plane, added to that of the rocket, gave reasonable penetration. Relatively heavy rifled guns were mounted on planes,
but they were a bit of an anachronism and never important. On
the other hand, the mechanism for firing a five-inch rocket was
merely a thin light tube, which added little weight beyond that
of the projectile itself, and a five-inch rocket shell would destroy
a tank in short order. Rockets even up to eleven inches in
diameter, the Tiny Tim that helped to polish off the last of
the Japanese fleet, were mounted on planes. It takes quite a ship
to mount eleven-inch guns. Of course the rockets were short-
range affairs, but the plane took care of that.

The air-borne-infantry assault should also be considered in
any discussion of improvement of defensive means on land, for
this might be a way of breaking defensive lines. Techniques for
air-borne assault were much improved during the war. The para-
chute and the towed glider were both made more dependable.
Light portable equipment of all sorts was developed. Signal
systems by which dispersed personnel could effectively gather
at night were given extensive attention. But the experience of
Crete, Normandy, Arnhem, the Rhine crossings, showed rather
well that the air-borne assault is an auxiliary to the advance
rather than a means of starting wide sweeps behind the enemy
lines. Wide cavalry sweeps, in these modern days, require
vehicles and plenty of them; any slow caravan in an armed
hostile country will be overcome by highly mobile forces concentrated against it. A weak island might be thus attacked, but

it would have to be decidedly weak. The much heralded air-borne assault is an aid to a break-through and apparently not much else. We are not interested in, and do not consider of course, the assault by an armed nation upon one that is un-armed in a modern sense—here air assault might work, but probably would not be necessary. There might be a place for air-borne assault in connection with surprise invasion on the scale of an enlarged Pearl Harbor, although this is doubtful. As an aid to break-through, to reduce casualties and risk in any case where there is control of the air and when the break-through is going to occur in all probability anyway, the air-borne assault has its place, and as such an aid it was skillfully used. An air-borne assault attempting to penetrate a line adequately protected by early warning and interception radar, and possessing fast inter-ceptor planes equipped with radar for night use, would be a suicide attempt.

There is a strong indication, therefore, that the defense may again be in the ascendant in land warfare, that the deadlock of the first war might well reappear if antagonists substantially equal from the standpoint of skilled use of ample technical de-vices met again at a long land frontier. This might merely throw the weight of the effort elsewhere, but it is an important factor to consider as we look to the future.

Amphibious warfare is another matter. Certainly we can be sure that if neither antagonist could penetrate the enemy lines on land, neither could do so by amphibious methods on a thoroughly defended coast. But there are other conditions than those in which great armies, fully equipped with modern methods and fully trained in their use, face each other across a line on land. Surprise is still an element of war. The defense that can be concentrated at a small outpost is limited. Hence we need to consider amphibious warfare.

The peculiar conditions of the past war produced a notable development in devices for amphibious warfare. When one sends troops ashore he must either do so on friendly soil or penetrate enemy defenses at the beach line. In addition to all the usual difficulties of penetrating defended lines there appear the difficulties of getting ashore. As an offset, the high mobility of sea transport affords an element of surprise. Nevertheless, when we employed amphibious movements, we did so after we had at least temporarily overcome the submarine, where we had un-disputed control of the surface of the sea, under conditions where we enjoyed a large measure of air control, and with the aid of numerous powerful auxiliary devices. At Tarawa, where we tried the old form of assault, preparation by naval guns and nearly conventional methods of landing, the effort showed clearly that no amount of valor and doggedness, in the absence of modern methods, could overwhelm a prepared enemy at the beach without prohibitive casualties.

One of the most important improvements for assault was the rocket ship. For the preparatory bombardment of an area at short range, where precision is not necessary, the rocket, fired from special ships or from landing craft, can lay down a far more severe blanket than can be produced by guns by the same effort. Another element is air preparation, carried more deeply, guided by radar navigation and sighting devices. Naval bombardment also has its place in this combination, to take out by pinpoint fire the emplaced enemy artillery that would make the operation of rocket ships, with their short range, impossible. Blanket bombardment of small areas, with perfect hitting of strong points, is a powerful combination, but blanket bombard-ment of large areas is questionable at best. Hence the rocket of relatively long range and low precision is not of much use here, and probably not elsewhere. When one adds range to a true rocket he soon loses most of the advantages of weight and cost, and when he adds precision by guiding devices he pays heavily for the precision attained.

Landing craft took many forms, and we do not need to go into them. Very effective was a vehicle driven both in water and on land by tracks similar to but lighter than those of a tank, and armored only against rifles and machine guns, which could propel itself over outlying bars or reefs and up onto the beach. With this the assault troops had much more of a chance, if artillery had been silenced, than with boats.

After the first wave was ashore there appeared the necessity of prompt movement to their support of tanks and guns, fuel and ammunition. Ships that could force themselves against the beach and disgorge all these things through doors in their bows were necessary elements and functioned well.

Then, when troops were established ashore, came supply, in thousands of tons a day, reinforcements, food, ammunition, evacuation of wounded, all the complex paraphernalia of attack. For these needs, to secure surprise, artificial harbors were constructed on the coast of Normandy, although one of them was destroyed by storm and most of the supply ultimately went in over the beaches. Pipelines for fuel were laid across the English Channel, with admirable ingenuity, and finally all across France. Trucks, railroad cars, and locomotives were landed as the assault moved forward and when harbors had been secured, although it was demonstrated for the first time that harbors were not essential in the early phases.

One of the most interesting devices was the Duck, also spelled Dukw from the classification initials from which its name was derived. This was an army truck supplied with a tight body for buoyancy and with a propeller. The idea of the Duck sounds obvious, but it was not. It could run at full speed ashore; it could negotiate soft sand on the beach because the driver could reduce its tire pressure while operating it; it could maneuver about a ship and take a load directly from its booms; and, most surprising, it could land through surf. It could do this latter because it always had traction, even when it hit bottom, and hence did not broach broadside on. It could be handled and maintained by truck crews with very little, but highly essential, additional training. It came into use because of the vision and persistence of a small group of civilian engineers, plus the encouragement of unconventional generals with a flair for pioneering, and in spite of general indifference or preoccupation elsewhere in the services. In fact, there was probably more obtuse resistance to this device than to any other in the war. Its history shows both the dangers of channeling in war the development of new devices as strictly military matters need to be channeled in order to function at all under the prevailing chaos and also the benefits of giving progressive groups their heads.

The initial skepticism encountered by the Duck in military circles was eventually overcome. During one of its experimental trials on Cape Cod, a Coast Guard boat with seven men went ashore on Peaked Hill Bar at one o'clock in the morning in a forty-mile-an-hour wind. Coast Guard teams could neither launch a boat through the heavy surf nor approach the disabled vessel by water from the Provincetown side. One of our engineering crews, in one of the first models of the Duck, rescued the stranded men with ease and dispatch and, I was happy to learn, with photographs. By next morning, the disabled boat had disappeared. But I had the pictures of what was certainly the first rescue in history of a naval vessel by an army truck. I gave them to Secretary Stimson, who showed them, at a Cabinet meeting, to President Roosevelt and Colonel Frank Knox, then Secretary of the Navy. From that moment on, we found less resistance to this newfangled and strange contraption.

When the Duck began to perform in the field, it was thoroughly appreciated. It was used by commanders generally, who of course had plenty on their minds besides development, and used to great advantage, in Sicily, Italy, and finally Normandy, as well as the Pacific. It should be emphasized here that no criticism of individuals is intended, nor is there where other delays or lacks may be mentioned, but merely comments on a system of organization, about which more will be said later. In any case, the Duck arrived and performed mightily. We made landings that would have been disastrous without it.

Any amount could be written on other devices of amphibious warfare, navigation and signaling devices, means for removing underwater obstacles, and the like, and it is an interesting story, but not essential here.

The important point is that for a landing on hostile shores, the difficulties of the landing itself are superposed upon those of penetrating a prepared line and exploiting the break-through. Even if there is no line at all, a successful landing involves dominance in the sea and air and suppression of the enemy under the sea. Conversely, with all these, it is doubtful if a fully prepared line in the modern sense can be penetrated at all with tolerable casualties or even without limit on these. Whether the deadlock on land again appears depends, of course, on how current technical developments turn out; it depends also upon the question of the possibility of attaining air dominance, which we will discuss later; but it certainly looms as a distinct and rather welcome possibility. We need not fear invasion of our shores in the foreseeable future, if we are armed and ready. There are other means of attack besides invasion, and we shall consider these. But invasion is one aspect of the possible war of the future that need not produce terror in us now.

When we consider all that was involved, the successful return to the Continent in 1944 takes on new significance. It could not have occurred if we had not had the finest sort of collaboration among allies, if we had not had for years salutary partnership among military men and scientists and engineers, here on this continent and in Great Britain. It could not have occurred without fighting men of courage, conviction, doggedness, and skill. It could not have occurred without an extensive, effective industrial system from all points of view, with skilled, loyal labor and effective management. It could not have occurred in the absence of the inherent limitations of dictatorships and the small stature of Hitler. It could not have occurred without the ablest set of military leaders that the world has ever produced.

CHAPTER V

ON THE SURFACE OF
THE SEA

"The United States had to learn the hard way. But we learned. We learned that team play brought success. We learned that any military effort of consequence required unified control to exploit the maximum capabilities of ground, sea, and air forces."
                                                            —ROBERT P. PATTERSON

THE LAST CONFLICT may well have seen the end of that most dramatic of the scenes of war: the clash of great fleets and great ships.

The whole evolution of sea warfare in World War II, from a technical standpoint, revolved about radar, a word that is a contraction of the phrase: "radio detection and ranging." Radar had its real origin in 1925, when two physicists performing experiments for purely scientific reasons sent out short radio pulses and studied their reflections from the ionosphere, the conducting layers above the earth that reflect radio waves. There are two ways of using radiation to find the distance to an object; one, to examine the interference pattern of continuous radiation, and the other, to send a short pulse of radiation and examine the echoes. The first was used in the proximity fuze, and is excellent for short distances. The second readily won out as the better for naval use.

Radar was one of the two really important instruments of war developed during the peace. The other was sonar, which uses the pulses of sound waves under water. In Great Britain, and also in the Army and Navy in this country, progressive officers and civilians, working in government laboratories and with industry, reduced the basic ideas of radar to workable form before the war started. In Britain particularly, a small group that realized the possibilities kept the pressure on, so that, when Hitler's air assault on Britain began, the intrepid air fighters were multiplied in effectiveness by radar guidance, and hence were able to down and turn back the fleets of bombers. These pioneers, scientists and engineers, undoubtedly are members of that few who earned Churchill's immortal tribute.

After the war was well under way, scientific interchange between the United States and the United Kingdom was brought about, but not without difficulty. An amusing aspect, if anything so obtuse and serious can be amusing, is that there were groups in both countries who opposed the interchange on the ground that they would be giving too much, and both were thinking of radar and both thought they had a monopoly. When interchange finally occurred, it soon appeared that both parties gained immensely by the combination, for the individual strengths and later the joint strength were enhanced significantly; and it also appeared that British science contributed more, for it furnished the magnetron.

The magnetron became the heart of radar. Strangely enough, its basic form appears in old German patents. But the Germans lagged on radar throughout the war, and fortunately the allied team, with Canada joined in, cut rings around them. The magnetron made it possible to produce pulses of very large power, but of short duration—a millionth of a second more or less—so that the average power was small. Combined with exceedingly sensitive receivers, and developed by a whole new theory that fills many volumes, radar came into its own. At the beginning of the war, radar sets were of relatively long wave length. It soon appeared that short-wave sets—centimeter radar —had the edge, and newly formed groups concentrated on that. Even after they captured some of this equipment, the Germans could not even copy it effectively.

The sensitivity of radar, and its precision, are both remarkable. The reflection of a pulse from an airplane two hundred

miles away can be detected, and its echo separated from the
thundering echoes coming back from near objects, even though
the same plane could not be seen by eyes against a mountain
background more than a mile away. Range is limited only by
the horizon, but it is severely limited thereby, for the paths that
the pulses follow either do not bend or at least bend very little.
Timing of the interval between the pulse and the echo can be
made so precise that the distance to an object miles away may
be measured within a few feet, and the pulses fly with the speed
of light. Using cathode-ray tubes to display the pattern of returning echoes, in much the same way that a television tube
displays a scene, radar can make the earth visible from a plane
above the clouds, in fog, or at night, with a detail as though
one had a map spread out before him or could see through the
obscuration. Such an instrumentality was bound to revolutionize many departments of warfare, and it did, especially on the
sea.

The mechanisms for aiming the great guns of fighting ships
were complex and effective long before the war began. The
first salvos from the *Bismarck* sank the *Hood* when the latter
was just a faint spot on the horizon. When it was possible to
aim just as precisely by means of a dot on a radar screen, sea
warfare took a new form. The saying was still heard in the terms
of rugged seamanship that we must be ready to fight with or
without radar, but he who fought without fought blinded. We
lost a battle—first Savo Island—to learn this lesson, but we
learned it.

There were great fleets of fighting ships in existence when the
war started and they were bound to clash. There is no need to
review the engagements, although many were examples of
masterly tactics and others were not. As always, in naval shooting war, the decision in the long run went to the ships that could
shoot straightest, soonest, and under most difficult conditions.
After stumbling beginnings, we could shoot best, for we had
the best radar, and the issue was no longer in doubt. The days
of great battleships seem to be over; we would perhaps use

them if we had them, but we would not build them. Their vulnerability to land-based air attack was shown at Malaya and in other places, and their protection against modern submarines has become more difficult, so that there is now serious question whether their value for any purpose whatever is worth the great cost of defending them. They passed from the scene because the great carrier took over their functions, with its longer striking range. Even the day of the great carrier may be past. It is by no means an open-and-shut question, but one that will stand plenty of dispassionate analysis. We shall see.

Into the functioning of the carrier, radar enters at every point. Something will be written later about protection against underseas attack. But the carrier, like every other ship that comes within the range of enemy aircraft, sea- or land-based, must protect itself against attack from the air. This need for protection is of three kinds: against low-flying planes, against planes at moderate altitudes, and against planes at great altitudes beyond the range of the ship's guns. The second was the usual problem early in the war. Here radar and the proximity fuze made a powerful combination. Automatic guns can be made to follow the indication given by a radar set and to point precisely at all times. Electronic computers can give proper lead for the speed of the target, allow for all the vagaries of the trajectory, cancel out the effect of roll of the ship, and in general aim with a prescience that verges on the uncanny. The guns can be caused to load and fire themselves at a high rate. Finally, proximity fuzes can trigger off the shells at just the right instant for most lethal effect. For a single plane or a few planes at moderate altitude to attack an alert ship thus equipped is suicide.

But a great ship is a valuable target, worth the sacrifice of planes that come in droves. These can hug the sea to avoid long-range radar until late in the attack. Kamikaze, or suicide, pilots can dive to death, hoping to carry the ship with them. Simultaneous attacks from many directions can saturate the main antiaircraft batteries, and some attackers may then slip through

for close assault. Thus there were needed secondary antiaircraft batteries, employing rapid firing at short range, for close-in
protection. These small guns, with gyro-lead computers, and
tracers, did not depend on radar, but they could put up an
enormous volume of fire. Thus protected, ships could still move
ahead in the Mediterranean and the Pacific in spite of aircraft,
and the great ship could in general defend itself.

When planes took off from a carrier to fight an enemy carrier,
radar entered throughout. Their own radar found the enemy,
and when they returned from attacking it they could be unerringly vectored back to a safe landing by their own carrier's
radar. There could thus even be night carrier operations. Radar
bombsights could also be used, but they had their principal use
elsewhere.

Could not radar be jammed to end all this? Of course it could
be jammed, and enormous effort was put into jamming or foxing
it in a hundred ingenious ways. But there are limitations. If a
ship starts jamming, it renders itself highly conspicuous, for
the act of jamming is to send out a terrific noise, electrical of
course rather than acoustical, but none the less revealing. As
between ship and plane, the former can carry more weighty and
intricate apparatus, use various frequencies and the like, and
the corresponding limitations render it unlikely that a plane will
often jam a ship. The art of jamming found its greatest field
in the air over Europe. At sea it did not enter strongly in fights
between ships and planes.

When it came to land-based air attacks on shipping, radar
found one of its most interesting applications. This lacked major
importance merely because the Nazis, when they had the
targets and the aircraft, did not have the effective radar, while
we generally lacked targets. It proved possible to construct a
radar bombsight of such precision that a plane flying at night
at relatively low altitude could almost drop a bomb down the
smokestack of an enemy merchant ship that was proceeding
without radar warning or antiaircraft defense effective at night.
In the Pacific seas, where only Japanese ships were encountered

and identification was not needed, one air squadron scored hits and sinkings with two thirds of the bombs it dropped. Unprotected shipping can no longer move within ready range of enemy land-based aircraft equipped with proper radar.

The war ended, or became a contest against a relatively unskilled or unequipped enemy, before the full evolution of warfare on the surface of the sea had run its course. So one development of great potential significance that appeared here and there did not come into full use. It is well worth examination, for it may decidedly alter the course of future events. This was the guided or homing bomb.

Bombing from moderate or high altitudes with ordinary bombs is a very haphazard affair. In spite of the aura of mystery drawn about the Norden bombsight, in spite of the bizarre claims of some overexuberant air-power enthusiasts, the fact remains that if half the bombs launched at a target dropped within a thousand feet of it, this was usually considered good performance from high altitude. A ship on the sea makes a clear target, but bombing with dispersion of that order will not hit it often, and bombers will not make a leisurely bombing run at relatively low altitude if the ship has any effective antiaircraft protection whatever, let alone the kind that now is possible.

It is quite possible, however, to guide a bomb after it is released, and throughout the half minute or so of its fall, and thus attain real accuracy. This can be done either automatically or with manual controls, and it was done in diverse ways. With a manual control, the bombardier sits after bomb release with a small joystick in his hands; as he moves this, so the bomb moves, for plane and bomb are connected by radio to transmit his signals, and the bomb has tail surfaces to steer it, and a gyro to keep it from becoming confused as to direction. He can either watch the actual ship and the actual bomb, the latter having a flare in its tail, or he can watch two radar pips on a screen that correspond to the ship and the bomb, and a simple computer can help his judgment. In the automatic form all this is done by a mechanism inside the bomb itself, operating by radar or by thermal controls. There was even one form in which the bomb carried a television transmitter in its nose, but this verged on the warfare of Buck Rogers or Flash Gordon.

This sort of device first appeared in the Mediterranean during the Italian campaign, and the Nazis introduced it. This probably was a case of introducing a novel device prematurely, before sound engineering had shaken it into reliable form, for which Hitler's well-known intuition probably was responsible. The homing bomb in question was of a type that, theoretically, was easily jammed, but it was so erratic in performance that, after jamming came into effect against it, there was long doubt whether it was being jammed or just being contrary. In fact, after the war, the Germans claimed that it was not jammed at all, and that for such an eventuality they had ready a bomb that as it fell unreeled a wire along which controlling impulses could be sent, but that they never used it. It is more probable that the jamming was a bit subtle, and that the Germans themselves, when their bombs flew all over the sky, could not distinguish between jamming and faulty mechanism. Nevertheless, the basic concept of the bomb they did use was sound and the weapon decidedly formidable. One wonders what would have happened if it had been really perfected and then put into use as a surprise and in ample quantity. As it was, it sank ships and was decidedly disagreeable to meet.

We introduced several forms of guided bombs, but none of them under conditions that gave extensive experience. In Burma, where an air effort to take out important bridges had been under way for a long time without much success, controlled bombs were introduced and took them out promptly, the records indicating that one controlled bomb was worth one hundred ordinary ones. But most of the forms came too late to be of much use; in the European theater they did not catch on, largely because no really precision bombing was being done or attempted there except in support of troops, where rocket-equipped planes did very well indeed. We are not concerned with whether guided bombs were used as much as they might have been used to advantage—perhaps they were under the circumstances that obtained, perhaps we missed an opportunity because of a type of conservatism that exists even in an air force. We are concerned principally with the future, and here guided bombs may be important indeed.

Bombing planes, it must be remembered, can be expected to fly above the altitude guns can reach effectively. This seems to be something we have to accept, for guns as we have known them have been pushed close to ultimate limits. The guided missile, which we shall later discuss, enters to affect this picture. Planes can fly fast, so fast that a pursuit plane launched from a ship at the instant when early radar warning is first available will have great difficulty indeed in climbing fast enough to reach the bomber before it comes to its bomb-release point. But bombers at very great altitudes and speeds have their limitations. They can seldom see a target on the ground clearly except by radar, especially if there are no strongly marked features. With ordinary bombs, which fly many miles horizontally as they drop, they cannot hit the side of a barn—they cannot even hit a small city with any assurance. The guided bomb alters this whole situation. It can hardly be made to hit with certainty a confused target such as one building in a city. But a great ship alone on the sea is a clear target to radar, and a clear target for a guided bomb.

Is there any effective defense against the very high bomber using guided bombs? If there is not, the days of large fighting ships—carriers as well as battleships—are over. We shall consider defense in connection with the evolution of bombing and of guided missiles. In all probability there is a defense, but how effective it may be will be questionable for a long time to come. This is not a matter to be judged hurriedly; one must remember that it takes years to build a large carrier, and that the decision not to build is the irreversible decision if war comes suddenly after an interval.

The primary mission of our Navy in war is to interrupt enemy sea commerce and to make it possible for our commerce to move safely to supply our allies and our fighting forces overseas. If there is a great enemy fleet in being, capable of interrupting our commerce, then the mission includes the destruction of that fleet. A secondary mission is to collaborate with the Army in amphibious operations for landings on hostile shores. During the past war the Navy carried out both missions with ultimate success.

In a possible future war, if it should occur while the world has its present general interrelationships, conditions would be far different. The only enemy fleet of any moment would be a submarine fleet. There would not be any enemy sea commerce of importance to interrupt. Our landing would be, we trust, on friendly shores, with amphibious operations secondary, if occurring at all. Certainly this will be true if we aid Europe to rearm and if there is time, and certainly to fail to do so would unnecessarily sacrifice enormous advantages. To maintain our lines of supply would be just as important as it was in the last two wars. In addition to submarine attack we might have to meet attacks upon our merchant ships by long-range, high-flying, land-based bombers searching by radar and employing guided bombs.

The mission of the Navy will be as important, and as difficult, as it has ever been in history. It will need to employ modern techniques to the utmost, and in ample quantity. But does it need great ships for its mission? Certainly until we have the means fully in hand for discharging the primary mission it would be foolhardy to seek out new tasks for great ships, such as participation in strategic bombing, merely for the sake of having great ships. Their cost is large, and their impregnability questionable. On the other hand, if there is an essential aspect of strategic bombing that can be effected only from carriers, and if they can be defended with reasonable effort and assurance, by all means let us build them before it is too late. There are two points here. First we need to analyze the actual situation intimately and without preconceived notions or disproportionate service loyalties. Second, the Navy has plenty to do to combat the submarine and protect shipping against enemy aircraft, and for this it needs ample strength.

The day of the great ship may be over. Never again may the world witness the moving spectacle of a mighty ship, manned by two thousand men in precise control of powerful mechanisms, moving irresistibly on its way into combat, or the sudden disappearance of its concentrated might in explosion and flame. Yet the days of the Navy are not over, nor are its missions less essential. We are a power in the world, and we intend to exert that power, if need be, far from our shores, to support our friends and strike an enemy where he is most vulnerable. It will involve new techniques, a new type of thought, new traditions. Mosquito craft of many sorts will be needed. Co-ordination of a new order, and in new ways, will be needed to combine air-borne and sea-borne weapons. We shall still need to sweep the enemy forces from the seas, whether they are under its surface or above it. We shall still need to make it possible for wallowing freighters, carrying essential heavy cargoes to our bases, our armies, or our friends, to move over the high seas with assurance. We shall need to be sure, we may need to demonstrate, that a world that depends for its prosperity upon sea-borne commerce cannot be torn apart by a land-confined aggressor, merely because some techniques of warfare on the surface of the sea have become obsolete and new methods are yet to be learned.

CHAPTER VI

WAR IN THE AIR

"Somewhere, at last, the limit of human endurance can be reached; but the voices of the surviving members of unconquered populations from Chungking to Coventry raise a chorus of denial that it is to be reached with any such ease as Douhet had foreseen." —EDWARD WARNER
Makers of Modern Strategy. 1943

STRATEGIC BOMBING, in which air power carries on warfare on its own, began with the Battle for Britain. The Nazis, having overrun the Continent, turned their great air fleets against England. We are not here concerned with the valor and doggedness of the British, alone and with their backs to the wall, but with the techniques involved in that battle. From this standpoint the tide was turned by two factors. First, an interceptor plane, if it could find an enemy bomber in the air, could usually bring it down. Second, early-warning radar was present.

One should expect, in general, that defensive aircraft would outfight the bomber. They need have only relatively short endurance, they are not hampered by a bomb load, and, most important, they operate over their own territory, where they can be guided or aided from the ground. They should be faster, with a higher rate of climb, and more maneuverable than the bomber. Great bombers are essentially fragile instruments, relatively lumbering in their flight, and dependent for their safety upon ignorance on the part of the enemy of their intentions or their positions. There were times later in the war when bombers could protect themselves, when found and attacked, merely by their combined volume of fire, but this situation, as we shall see, is not likely to occur again between equal adversaries. In general, bombers must sneak in and thus avoid the defensive fighters. This was certainly the case when the war opened.

Early-warning radar tipped the scales in that first great encounter, and it was available only because a few scientists and military men had been able to foresee what was coming and have their way. The defensive fighter fleets of England were sorely limited in number, they took heavy and almost fatal losses, but by radar they were enabled to meet the enemy, and they prevailed.

What ensued was a race in techniques. We can touch on only a few high spots and review the general pattern; and to do even this is difficult, for the instrumentalities were highly complex, and their interrelation taxed skill in organization and training to such a point that a detailed account would be extensive indeed. We examine in order to see trends, and shall omit much. We shall not attempt to examine the swaying course of conflict, as bombing of Britain gradually gave way to bombing of Germany, as the air fleets of the United States entered the fray, and defense mounted more effectively in Great Britain than on the Continent. There are two main methods of defense, by fighting aircraft and by antiaircraft artillery, and we must consider both.

The direct clash of fighter fleets played a great part. This would still be important in any case where adversaries faced each other at short range, as they did then, or in connection with land campaigns, but as a feature of strategic bombing at long range it is not likely again to appear. At short range, fighters could drive the enemy from the air, thus clear the way for the bombers, and then accompany them to take on the surviving defensive interceptors that rose against them. This combat of fighter planes furnished an analogy with the battles of knights, a few highly armed and mobile units, fighting it out on their own, and largely immune to the acts of the great masses below. But dogfighting in the air, as a determining feature of really modern war, is probably now a thing of the past.

The appearance of the jet fighter changed that. The jet engine is compact and has enormous power for small weight, but it is a voracious consumer of fuel. If jet engines could ultimately be made economical in fuel consumption, that gain would emphasize the change that was produced by their advent. Jet engines increased the disproportion between the performance of the long-range bomber and the short-duration fighter. They made speeds in the air so high and turning radii consequently so large that dogfights became almost impossible. To conduct a dogfight, one has to be able to keep the enemy in sight for more than fleeting seconds. Jet engines increased the effectiveness of the fighters' operation over their own territory with the aid of land facilities, and correspondingly decreased the possibility of defending bomber fleets by accompanying fighters. They increased the influence of mechanization in the air and reduced the premium on human skill, for flying alone approached the end of the possibilities of unaided senses. They may well have made mass bombing at moderate altitudes, against a fully prepared and alert enemy, obsolete.

To understand this one needs to review the radar means employed in defense. First there is the long-range early-warning installation, located on coasts or borders, capable of detecting planes as far as the curvature of the earth permits—at two hundred miles or so if the incoming planes are high. Second is the interceptor-control radar. This places on the cathode-ray screen images of all the planes within its range in the sky. It reveals not only their positions but also their altitudes. It can also contain means for identifying friend or foe. Its task is to control the movements of interceptor planes by radio communication, so as to place them in such position as to attack the enemy to greatest advantage. Finally, there is the radar on the planes themselves, necessarily of relatively short range because of weight limitations, used to effect the final contact, and even to fire the guns, and available also, of course, to the bombers for defense. There are also the radar and other aids at landing fields—ground-controlled landing systems and the like—to bring fighters or bombers down safely on their darkened fields.

To penetrate a defense thus equipped, night or day, if it is working smoothly and there is enough of it, is an appalling task for any bomber fleet. There are three general methods of doing so—reduction, jamming, and sneak raids. As our strategic bombing offensive against Germany developed, the fleets of Great Britain operating principally at night and ours principally by day, we depended upon all three. The enemy fighters were to some extent driven from the air, either by direct combat by accompanying fighters or by loss of fuel and production facilities under the continued bombardment. All sorts of jamming and deception tricks were employed. The versatile Mosquito planes of Britain turned in a great performance by slipping in at low altitude below effective radar detection. Yet the contest was touch and go even before the German jet plane appeared, and it would have been prohibitively expensive if this fast, able fighter had appeared in large numbers, backed by adequate radar, before Germany collapsed.

Complex navigating systems for aircraft were developed. The long-range loran system for this purpose, using reflection from the ionosphere, covered the oceans. In this system, simultaneous pulses of radiation were issued from three or more stations, and their times of arrival at plane or ship were compared by a suitable cathode-ray device, yielding the position rapidly and within a mile or two. Systems more precise than loran, but operating on the same general principle, were in use for guiding bombers over the Continent to their targets. Pathfinder planes at night, using these aids and radar beacons and sights in addition, led the fleets and placed flares to guide them to the target.

Enemy antiaircraft batteries were radar-controlled as well as our own, but the Germans did not perfect microwave radar. Jamming took on great importance and many forms. The straightforward method of jamming is to carry on the plane an electrical noise, or rather a maker of electrical noise. If this is operating at the frequency of the radar set, it can fill the set's screen with confused patterns and drown out the weak echoes on which the radar depends. But if the radar set is versatile, it can avoid this trouble by shifting frequencies. A noisemaker capable of sweeping whole bands of frequencies can be built, but it will be either weak at a particular frequency or heavy to carry. Another way of confusing enemy radar is to emit fake echoes, to give the impression of planes that are not there, or of planes in the wrong place. To do so requires knowledge of the frequency being used on the ground, and such knowledge can be obtained in the air. But all these things have severe limitations if the enemy is in a position to use many frequencies and many sets of radar, particularly in the centimeter range. A more general deceptive means was "window," the stratagem by which clouds of sliced aluminum foil are released by a plane to give masses of echoes and cloud the radar screen. The distributing plane itself is likely to be a bit conspicuous, but the method was highly effective at times.

Antiaircraft artillery, radar-controlled and unjammed, with advanced types of computers and proximity-fuzed shells, can stop any fleets of bombers that fly over it within its range. If its radar is fully versatile, more so than the types used during the war, employing as one artifice means for distinguishing moving objects and ignoring stationary ones, it is more difficult to jam than conventional radars. Fortunately, the Nazis lagged in radar throughout the war, largely because they expected a short war and were dominated by obtuse and dictatorial leaders who put their young scientists and technicians into the ranks. Fortunately again, the Nazis never developed a proximity fuze.

They tried hard enough. They knew well enough that such a fuze could multiply the effectiveness of antiaircraft guns by a factor of five or ten, and recognized when their turn came to be bombed that such a fuze might well tip the scale. When their massed fighters met our bombers at Schweinfurt, and spread havoc among them by a frontal attack with rockets, examination of fragments showed that the rockets were supplied with recesses to receive proximity fuzes. Those recesses were never filled. If they had been, and if the same development had been successful in shells, we would have been stopped cold on that type of bombing. As it was, we proceeded only when, by means of discardable fuel tanks, the range of our fighters was increased sufficiently for them to accompany the bomber fleets. The history of Nazi failure with proximity fuzes is an interesting one. When we introduced the fuzes in artillery we had teams of men at the front to detect the first sign of enemy jamming of the devices, and these men were ready with the next step. But jamming never came, in the five months that remained of the war. The Nazis just could not believe that the *verdammter Amerikaner* had produced where they had failed.

In the air rockets were important principally in air-against-tank combat. The rockets fired against aircraft in defense of Britain made a lot of noise, but accomplished little else, although even noise may have been of some comfort when the bombs were dropping. Rockets employed in interceptor planes never superseded the concentrated fire of a group of machine guns for air-to-air combat. By the time we had installed proximity fuzes in rockets we had few targets. But this fact does not mean that a rocket thus equipped does not provide a powerful weapon. It will be used in air-to-air combat in the future unless the guided missile forms a still more powerful weapon; we shall discuss that later.

By the time we operated fleets of bombers against Japan, enemy defense in the air was weak. Japan did not proceed far along any important technical lines. Her military parceled out small problems to her scientists but kept them severely at arm's length. There is nothing wrong with the Japanese mentality when it comes to fundamental science; Japan has produced some notable scientific figures and is beginning to do so again. But her system was wrong, and she got nowhere. Our fleets of B-29's, in mass formation, with their enormous firepower arranged to bring concentration to bear against any attacker, were well able to take care of themselves. They were correct for their purpose. They would still be correct against any target similarly weakly defended. But no fleets of bombers will proceed unmolested against any enemy that can bring properly equipped jet pursuit ships against them in numbers, aided by effective ground radar, and equipped with rockets or guided air-to-air missiles armed with proximity fuzes.

All this discussion of defense against bombers is pertinent also in connection with possible deadlock on land. It is doubtful if fully defended lines can be penetrated in the future, and still more doubtful unless one side secures air dominance. The trends, especially because of the advent of the jet fighter, are against this in the long run. The use of heavy bombing fleets, at relatively low altitude, to aid in breaking a defended line, looks to be very dubious in the presence of plenty of enemy jet pursuits. Bombing a country by conventional means so thoroughly that no jet pursuits can be maintained also looks dubious. Even with the atomic bomb it may well appear that any attainment of air supremacy to an extent sufficient to break land lines must also be so overwhelming as to render breaking them unnecessary.

The next point to consider is evidently the bomber that flies high. First, the argument runs, if the bomber flies high enough it can be beyond the range of guns. This is not an open-and-shut matter by any means. The high-altitude bomber with its pressurized cabin is decidedly vulnerable if it is hit at all. Yet, though the range of guns is increasing, and at least those engaged in developing guns are by no means ready to concede that they have neared their limit, it looks as though the importance of antiaircraft artillery of the conventional sort will decrease, even though it will have its uses, and even though it has become exceedingly deadly where it can reach. Second, if the bomber plane flies high enough, it eludes the pursuit ship unless the latter has very early warning to enable it to climb to altitude and engage. This depends of course upon how far-flung an enemy warning net is available and how fast fighters climb. The concept of immune fleets of bombers is undoubtedly a sound concept from a temporary standpoint, but the matter is not so simple as that of merely making a bomber of high ceiling.

In the first place, the high bomber has its difficulties when it comes to hitting land targets. It will almost necessarily depend upon radar sighting, and this is useful only where they are prominent features such as lakes or rivers to show up conspicuously on the radar screen. But radar sights can be jammed and deceived, and visual sighting is not much good, even when it is daylight and there are no clouds.

In regard to jamming of radar sighting devices, there will undoubtedly be a complex story. For a ship to jam makes that ship conspicuous, although of course one ship can jam and protect another. But on land there is no great harm in having a conspicuous point near a protected area. The attacking bomber will give away its radar frequency as soon as it opens up its radar, and it must open up some time before it releases its bombs. This gives the jamming station its chance. The bomber, moreover, can hardly carry the weight to afford a highly complex radar set with frequency range and other advantages. Just a bit of jamming may confuse a radar sight, while much more is needed to obliterate an explicit blob on a screen. There is no weight limit for the jamming apparatus on the ground.

Bombers may fly high, but they may not be able to hit much, at least with conventional bombs. They may use directed or homing bombs to help them out of their difficulties. Against ships these are likely to be of great moment, in spite of the use of jamming, false targets, shields in the form of surrounding antiaircraft vessels, and the like; a long and complex contest between the means of attack and of defense can therefore be foreseen. One hit with a really large bomb is sufficient to dispose of a ship, whereas many such bombs are needed for a land target of any size. The atomic bomb forms a special case, which we shall discuss later. But in the use of conventional bombs against land targets from very high altitudes, guided and homing bombs, while they will aid, will not be likely to overcome the disabilities of flying very high. Homing bombs may be decoyed to the wrong places, if one has area in which to do so. Manually guided bombs are useless unless the bombardier can see the target. If he seeks to do so visually, and even if it is daytime and there are no clouds, the efficient smoke producers developed during the war, which held seventy miles of the Rhine under smoke for days, can conceal a city. It will do him no good merely to put a television transmitter in the nose of his bomb to extend his vision in such circumstances. If he relies on radar to allow him to see the land below, this can be jammed.

We have one more technique to consider. Even without it there is great question whether bombing fleets, low or high, can ever again operate successfully with conventional bombs against a fully defended city having all the resources of protection available and alert. There is still more question whether the game would be worth the candle. The additional technique we have to consider is that of the guided missile. This will be discussed in detail later. Here, it suffices to say that, with the guided missile developed to its potentially attainable deadly form some time in the future, it may not even be feasible to carry many atomic bombs successfully to fully defended and alert targets. At the present time strategic bombing is an important element in possible war. With the atomic bomb available it takes on new aspects, which we must consider. But bombing as we have known it, and as our friends have suffered from it in the recent past, may indeed be obsolescent. The day may be approaching rapidly, if it is not already close upon us, when great fleets of bombers, at high altitude, carrying conventional bombs against a prepared adversary, are not a warranted or justifiable undertaking in war.

If we leave the atomic bomb out of consideration for the moment, we may well conclude that the defense in war is again approaching ascendancy. Deadlock on land is entirely possible. The clash of great fleets at sea is not again in sight in present circumstances. Conventional mass bombing may be obsolete.

This would be interesting news indeed, and salutary for a weary world. But we cannot come to the comforting conclusion that there is a potential technical deadlock between fully prepared nations as yet. The atomic bomb enters. We also have yet to consider the war under the sea. Here the attack seems to be in the ascendant, and to this we come next.

CHAPTER VII

UNDER THE SEA

"Captured U-boats alongside a New England dock may look drab and scantily-equipped compared with our latest models. Yet these ships came within a handsbreadth of mastery of the world. Thanks to them and to their like, Germany, till past the spring of 1943, seemed to hold victory in the hollow of her hand. Of what avail to build up an American Army of eight million men if it could not be transported across the Atlantic to come to grips with our principal foe?"

—JAMES P. BAXTER, 3RD
Scientists Against Time. 1946

WE HAVE TWICE ENTERED war while underestimating the power of the submarine, and twice the outcome has been in doubt. We must not do it again.

The first time, the power of the submarine came as a surprise, for it had just matured into a reliable device. This last time, we owe our mistake to overconfidence in an excellent antisubmarine weapon developed during the peace.

The first war showed the ability of the submarine to interrupt commerce, even though the U-boats of those days were clumsy, fragile affairs. They could make only a few knots, they were designed to submerge to a depth of only one hundred feet or so, and their hulls were so weak that a depth-charge explosion seventy-five or one hundred feet away would destroy them. At the beginning of the war there was no means whatever of detecting them while submerged, there were no aircraft to sweep the seas in search of them, and they could roam more or less at will. They were finally overcome by depth charges and the convoy system, aided to some extent by the beginnings of the whole art of detection.

These beginnings were crude. The first listening devices were exceedingly simple—a sort of stethoscope hung over the side, with rubber bulbs on one end and earpieces on the other. From a stationary craft one could listen with this to the propellers of the submarine and obtain a rough indication of direction. Before the First World War ended we had developed a similar device built into the search ship, capable of being used only at rather slow speeds, but much more precise in direction. With it a group of surface ships could sometimes follow a submerged submarine with sufficient effectiveness so that one of them could run ahead, drop its depth charges, and secure a kill.

The convoy system was merely a procedure of sending merchant ships in groups, guarded by antisubmarine craft equipped for search and attack. This innovation and a liberal use of depth charges substantially terminated the depredations, but not until after a severe struggle that might have ended disastrously.

In the interval of peace two things happened, one to the advantage of the submarine, one to its disadvantage. The submarine acquired more underwater speed, longer endurance, and a tougher hull. By the time the Second World War opened it could submerge rapidly to several hundred feet and maneuver in tight circles. The submarine could always use listening devices just as well as a surface craft could, and this fact, added to better performance, made the old forms of attack on the submarine practically obsolete. The depth charge, even when the second war began, was nothing but an ash can filled with TNT with a hydrostatic fuze to explode it at a given depth, and it sank at a rate of only five feet per second, so that the submarine had considerable time in which to get out of the way and, with its higher speed and greater depth of submergence, could usually do so. Moreover, the hulls were now so tough that nearly a direct hit was necessary for destruction. Finally, not only had the speed of submarines increased to the point where listening to them was less helpful, but in addition the submarines were much quieter than before. The stage was all set again for extensive depredations by submarines, which were again nearly immune to attack.

However, during the peace, a very important means of detection had been developed in naval laboratories both here and in England. This was called asdic at the time, and later sonar. It is to underwater detection what radar is to air. The ship carrying it emits a pulse of sound of relatively high frequency and then examines the pattern of returning echoes. By timing the interval, and by a directional receiver, the distance and direction of the submarine can be obtained. In later forms asdic or sonar could also supply a determination of depth of the enemy. It was a powerful adjunct indeed.

It is important to note that radar pulses cannot travel more than a few feet under water. Moreover, there are strict limits as to what can be done with sonar pulses and their reflections, because of attenuation of the signal and confusion by echoes from the bottom, so that a few thousand yards under good conditions seems to be the limit of range. There is no such thing under water as radar's range of many miles in air. Still, sonar seemed to be the answer to the submarine as it then existed.

Perhaps it would have been, if there had been an equal advance in attack weapons, but we entered the war with the same old ash cans and held to them tenaciously even though they were clearly outdated. The undisputed power of sonar gave us overconfidence, and we nearly came a cropper.

This was well shown after the second war got under way. Though the military services then began to accept the aid of the independent scientists and engineers of universities and industries in almost every other branch of technical warfare, they decidedly felt in this country that in the field of antisubmarine devices they had the matter fully in hand and needed no aid. This was a costly attitude before it gave way to close collaboration. It is entirely possible that some of the admirals had forgotten their courses in physics at the Academy, or were more interested in other things than the prosaic job of ridding the seas of underwater raiders. At any rate, those in charge of antisubmarine preparations were much too cocky, and we paid for this later in excessive sinkings.

Fortunately, secrecy had been well maintained during peace,

and this to some extent offset the lack of progress. Sonar devices had been brought into excellent form in government laboratories in collaboration with manufacturers, good training was instituted, and antisubmarine craft were equipped in numbers both here and in the United Kingdom and Canada. When our later allies began to take the full brunt of the Nazi submarine campaign, the Nazis were ignorant of the presence of sonar, and that ignorance cost them many a U-boat before they found out what they were up against. Otherwise, early sinkings would have been much higher. As it was, they were bitterly high.

Even the presence of sonar did not assure protection of convoys. Part of the trouble was the fact that sonar sometimes is severely limited, when temperature distribution in the sea is unfavorable; hence, much study was given to measuring and charting these conditions. Limits were set on the effectiveness of searching craft by the fact that their sonar would not work at high speeds. But the principal limitation was the fact that depth charges were not much good. Submarines could now work into position ahead of a convoy, discharge their torpedoes, and then by going to great depths be rather sure of getting away safely. Depth charges that were at least streamlined to sink more rapidly were belatedly developed, something was done about installing proximity fuzes on them, and these changes helped. But the odds were still on the submarine in surprise attack on a convoy or an isolated ship, and losses mounted to serious proportions.

A really formidable enemy of the submarine during the past war was the aircraft equipped with radar and armed with rockets. This is a decidedly deadly combination, whether the aircraft be one conducting shore patrol, or a long-range craft to go to the support of an imperiled convoy, or one based on an antisubmarine carrier.

The British learned this lesson early. Their Coastal Command rendered the waters about the British Isles almost safe, and the U-boat avoided them. We learned the lesson the hard way and took severe losses along our coast as a result. This was partly an organizational difficulty arising out of government by committee in the form of the Joint Chiefs of Staff—that is, a lack of effective unification in a war that demanded it—and partly a lag owing to congealed theory. These situations have been reviewed elsewhere and need not be dwelled on here. We finally learned, but it cost us much.

Another lesson we learned with great difficulty was the necessity of getting off the defensive and onto the offensive in the submarine war, of searching out the submarine and destroying it, of the proper use of hunter-killer groups for the purpose. It is an old principle of warfare that battles and campaigns are not won by remaining on the defensive. The defensive is necessary of course, in order to maintain a secure base from which to attack, but it is secondary only. Yet for a considerable part of the anti-submarine war we remained on the defensive, guarding our convoys with wavering success, taking heavy losses, and failing to employ to the utmost the new tools of war that made attack upon the submarine, even in the broad reaches of the sea, a feasible matter.

There probably never was a better example of the fallacy of leaving decisions entirely in military hands on the employment of new means of combat resulting from scientific and technical trends that they cannot fully understand. It is extremely dangerous to place military decision fully in the hands of brash amateurs overriding the judgment of professionals, as history has demonstrated; and the worst example of this is Hitler, with his intuition and gyrations. It is equally dangerous to allow scientific principles and trends to be judged by military men without review. We are living in a different world from that in which the needle gun was a notable technical advance. Yet the fact that the system during the early part of the last war was wrong and produced some mistakes was not in the last analysis the fault of military men. Under our system of democracy, the military are subordinate to civilians in government, even in time of war. If the system was wrong, it was the duty of civilians to correct it; the duty of the President, Congress, the civilian deputies of the President, and scientists and technical men who could see the dilemma most clearly. That a smoothly operating system was not produced during the turmoil of war is not surprising. What is somewhat surprising is that we have not yet fully corrected the fault. But the remedy is by no means obvious, and we shall speak of it again.

This chapter, in particular, must mention our faults and limitations, if we are to view the whole matter of war completely and objectively. There are those who delight in enlarging on the errors and misjudgments of others and take particular satisfaction when those they criticize so readily are military men carrying heavy responsibilities. Let it be clear that this review has no such point of view. Mistakes there were, and they were the fault of military and civilians alike. The serious mistakes were in fact few and far between; no war has ever been fought without them, and the record of this past war is remarkable in this regard. It is remarkable because we had military leaders of sagacity and resolution, because we had a dauntless war President, because such great statesmen as our wartime Secretary of War were effective in their support, and because the Congress of the United States, in many ways better adapted to peace, operated well in war. It should also be recorded that, while military men and scientists started poles apart and differed vigorously throughout, they succeeded finally in developing a partnership and a mutual respect that augur well for the future. We can return to the account of submarine warfare, errors and all, without apology for the overall record of the war.

The submarine war developed into a race between techniques, and for a long time the issue was in doubt, and with it the issue of the war as a whole. It was finally decided because the techniques of the Nazis were delayed in their introduction; they had too little and they were too late.

By late 1942 sinkings of surface ships had mounted to the point where they threatened to cut us off from Europe and terminate our land efforts. Only an extraordinary shipbuilding program was delaying the catastrophe, and this was straining the nation's resources. Worse than that, it appeared that the Nazis would soon introduce two new devices that might well determine the issue: the long-range torpedo and the schnörkel.

We know the outcome. These innovations were delayed. By the spring of 1943 the power of new antisubmarine devices became decisive; sinkings went down; the submarine was overcome; our supplies could move. We could stage the great invasion of Africa essentially without interruption. When the German advanced techniques finally appeared, the battle had become much too one-sided for them to restore the balance.

Let us examine what were these methods of attack that saved the situation. To be overcome, the submarine must first be spotted in the broad reaches of the sea, then more accurately located by pinpointing methods and pursued while it remains submerged, and finally destroyed. There were striking advances in all these fields.

There were two great methods of broad search, highly effective at the time, and now both obsolescent or greatly reduced in value. The first of these was direction finding by radio, and the reason it was of great value was that the Germans made an error that will probably not be repeated. They may have made it of necessity, because they did not dare to allow their submarines to operate on their own and out of control from Berlin. However that may be, they communicated by radio from Berlin with their submarines continuously throughout the war. Doing so enabled them, of course, to pass on information obtained in one way or another in regard to the position of convoys and to assemble their wolf packs for attack. But it was a boon to us. Radio stations listening to their messages could take cross bearings and give a rough position of the sending submarine. Consequently, we generally knew about where the enemy was located and could plan accordingly.

The second great method was that of radar. The submarines of the day had to spend a considerable fraction of their time on the surface to charge their batteries. When on the surface they could be detected by radar at a distance of twenty or thirty miles. A radar-equipped plane could therefore sweep great areas, as much as five thousand square miles an hour. Even the enormous areas of the open ocean begin to yield when fleets of planes are thus equipped.

This power was enhanced by new navigating systems, especially loran, which could give positions quickly and precisely. When submarines began to be encountered in the path of a convoy, long-range antisubmarine aircraft could be launched from the nearest land to sweep the seas and to add greatly to the convoy defenses; and the knowledge of exact positions, of convoy, plane, and detected U-boat, even when no celestial navigation was possible, added substantially to effectiveness.

The most deadly combination, however, was the hunter-killer group. A small aircraft carrier and other surface antisubmarine craft accompanied a convoy, or left it for direct attack upon a U-boat concentration. Radar sweeps gave the necessary knowledge of position, and specially equipped aircraft alone or in combination with surface craft pressed home the attack. When we thus carried the war to the enemy, the tide began to turn.

For precise following of a submerged submarine there were several developments. Sonar remained the mainstay, and, while its range was not radically increased, it became more versatile and dependable. There were countermeasures of course, and U-boats released fake sonar targets, but these were largely overcome, partly by ingenious recorders and partly by other ingenious devices that took advantage of the physical fact of a change of pitch in sound coming from a moving object. It finally came to the point where a submarine once caught in the sonar beam of a surface ship, when sonar conditions were good, could hardly expect to maneuver in such manner as to escape.

Magnetic air-borne detection, MAD for short, also appeared. It could find a submerged submarine by the distortion the submarine produced in the earth's magnetic field. The instrument was highly refined and sensitive, necessarily and always severely limited in range, the sort of thing that would seem to belong in a laboratory rather than in the rugged conditions of war at sea, and so regarded by some. Yet it could locate a fully submerged submarine from an aircraft, mark the spot, and follow the submarine, and this was and will be important. The device never came to full fruition and was used mostly from blimps in coast defense and in a few aircraft. Yet it very neatly denied the passage of the Straits of Gibraltar to submerged submarines proceeding quietly with the tide, after all other means had failed.

The sono buoy supplemented these means and was particularly useful from aircraft. One cannot listen to a submarine's underwater noise from an aircraft, or for that matter from a surface craft at high speed, which is itself making so much noise as to drown out all else. But one can drop a small buoy to lie quietly on the surface, listen through a microphone that it lowers to suitable depths, and broadcast by radio what it hears. The value lies in the fact that such a device may be made light and compact, capable of being dropped from an aircraft at full speed, and relatively inexpensive when produced in quantity by modern manufacture. When a submarine submerged after having been caught on the surface by radar, it was no longer safe, even if no surface-hunting ships were present. Aircraft could keep contact with it by rings of sono buoys, each dutifully sending its identification signal and the news of the submarine's progress. Aircraft, in relays if necessary, could thus keep the submarine under surveillance until its underwater endurance was exhausted and it was forced to surface and meet its pursuers, or until surface craft could arrive and take up the trail by sonar. Moreover, new means appeared for attacking it while it was still submerged.

The depth charge was really obsolete, but remained in fashion for a long time. Forward-thrown weapons were inherently more effective, but were a bit slow in becoming available—naturally enough, for it is not easy to make devices rugged enough to stand the battering of seas and the corrosion of salt water. The large depth charge lost its position in combat when submarine hulls became so thick that nearly a direct hit was necessary for lethal effect, so that many smaller charges designed for direct hits became more effective. The forward-throwing weapons, hedgehog and mousetrap, sprayed a pattern of charges well ahead of the attacking ship, by spigot guns or rockets. These charges sank rapidly, and when one hit the hull of a submarine it blasted out a hole bound to be lethal. The weapon was designed to cover an area with a pattern of charges like a shotgun pattern, leaving no gaps, and thus shooting became more deadly. Most important, it could be directed to one side or the other in accordance with last-second sonar observations and could be fired while the submarine was still under sonar contact, not after the attacking vessel in passing over the submarine had lost contact.

The rocket-carrying plane was appallingly effective. To the U-boat crew on deck at night this must have been really terrifying to contemplate. A high-speed plane, having located the submarine by radar, could come diving in at a moderate angle. It could switch on a blinding light for a final guidance or fire by radar alone. It had the firepower of a whole battery of field artillery in the few seconds of attack. Its shrapnel-equipped rockets could sweep the decks of the submarine and remove antiaircraft gunners, if there were any hardy enough to be manning the guns. But its most deadly weapon was a rocket with a solid head. This was aimed to hit well short of the submarine; it had a long, shallow underwater trajectory; when it hit a submarine it went clear through, leaving a gaping hole on both sides and chaos within. Few submarines fully caught by surprise on the surface by such a plane could hope to survive.

The matter was not all one-sided, of course. Submarines could carry radar to detect the approach of ships or aircraft, but these themselves made the submarine conspicuous, so that mere radio listening devices were more commonly used, sweeping frequencies to try to detect the plane on or near an expected wave length. In this way submarines could often listen to radar pulses from a pursuer at distances too great for the pursuer's radar to detect the submarine. False radar targets could also be liberated to float on the surface.

But it began to appear that the submarine, in the presence of the plane, radar, and other deadly devices for attack, was coming to be in the position of a school of surface fish with a flock of gulls overhead. As the new devices came into effect, as hardy crews in the air or on the sea learned to use them unerringly even in the stress of combat and the confusion of night, the importance of the submarine declined. It almost appeared that the day of the submarine was over, not only for the war but permanently. It might have been, except for two new and important improvements in submarines themselves. If they had come in early, as early as was readily possible, they would have swung the pendulum back, so far back as to change the whole course of the war and make its outcome doubtful. That they did not is owing to a factor best expressed in the words of Admiral Doenitz: "For some months past," he wrote in December, 1943, "the enemy has rendered the U-boat war ineffective. He has achieved this object, not through superior tactics or strategy, but through his superiority in the field of science; this finds its expression in the modern battle weapon—detection. By this means he has torn our sole offensive weapon in the war against the Anglo-Saxons from our hands. It is essential to victory that we make good our scientific disparity and thereby restore to the U-boat its fighting qualities."

These two innovations were the high-speed submarine equipped with schnörkel, and the long-range torpedo.

The Germans had designs of high-speed, relatively quiet submarines and built a few, now known all over the world, but they did not build them in quantity. Their importance lies in the fact that they can readily get into position for convoy attack and escape afterward. In fact, if they can outrun the surface anti-submarine craft, or travel at such speed that the latter's acoustic devices are greatly limited, they are much to be feared. They step up the pace at sea. During the past war we ran such ships as the *Queen Mary* safely without convoy because they were faster than the submarines of the enemy. If the submarine speed is the same as or greater than that of such ships, this method is ended. High surface speed for positioning and high underwater speed for attack and escape are in themselves sufficient to place us in a new era of underseas warfare. It would not be so serious if we could readily make a corresponding increase in the speed of displacement ships, but we cannot do so for an entire merchant marine at tolerable cost.

The schnörkel was, however, even more important than more speed. It appeared toward the end of the war and was troublesome, to say the least. It is merely a pipe arranged so that the submarine can run submerged on its engines, with only a small end of the pipe sticking out, like a swimmer breathing through a long straw. An ingenious contrivance prevents difficulty if the seas slop over the top of the pipe, by momentarily closing the pipe so that no water enters, and this is the real technical advance. In the brief intervals when the pipe is closed, the engines draw their air from the hold of the submarine, with some crew discomfort, but otherwise without reducing performance. Such an affair as a schnörkel is very hard to see at all, either by radar or by eye, when there is a bit of sea running. It does not make the submarine completely immune to radar search, but it decidedly decreases the effectiveness of such search. It is undoubtedly possible to build submarines that run on their engines when completely submerged, thus dispensing even with the schnörkel. This possibility has its limitations and disadvantages, as well as great promise, but even the schnörkel alone is something to conjure with, as we found out when toward the end of the war we tried to hunt down submarines thus equipped.

The long-range torpedo is much more important than it might seem at first glance. It introduces an element of surprise that is highly significant. The point is that the range of sonar against submarines is limited, and the range of a torpedo can be made to exceed it if one makes the torpedo large and expensive enough. Thus, if a surface ship can know of the presence of a submarine only by its sonar, and if the submarine can fire a deadly torpedo from well outside sonar range, the advantage seems to be entirely with the submarine, which can pick off the escorts of a convoy and then close in for the kill. The matter is not nearly so simple as this might indicate, however. This account, limited as it is to those things which are now generally available to open search and study, can hardly treat it completely.

The fact that emerges is that the day of the submarine is by no means over. If we entered a war soon, against a technically and industrially strong enemy, and if that enemy could effectively apply modern devices at sea, we should have the whole job of overcoming the submarine to do over again on a new and unattractive basis. Again we should face the severe threat that a nearly immune submarine fleet might determine the outcome of the war in favor of the enemy. Many of the successful methods of the last war are now obsolete against the truly modern submarine. There is no cure-all. The scientists will not suddenly pull a rabbit out of a hat to reverse all the trends. To cope with the situation we need to be alive to the dangers, not lulled into complacence by partial successes—to be forging ahead in a dozen fields of difficult technical effort.

The situation is by no means hopeless, of course. If a submarine can be pinned down by air cover, using sono buoys, MAD, and diverse other devices, it can most certainly be disposed of, and it will not matter whether or not it is a modern submarine of high speed and long submergence. We may have to adopt new tactics, perhaps, to clear and protect sea lanes rather than to convoy. We may have to depend more than in the past upon mines and upon blocking the submarine in its ports. More probably, we shall employ methods and devices not now fully visualized, and developed as experience with modern submarines may indicate. The job will be tough. It can certainly be done.

The submarine is one of our greatest potential enemies. We look forward to a world in which peaceful independent nations may trade freely across the seas to their mutual advantage. We look to rising standards of living and relative prosperity in many lands. With all such peaceful nations, governed so as to be responsive to the nobler aspirations of mankind, we would collaborate on a basis of equality. With them, in the days before

One World, we should expect to be able to maintain the peace
by combined force of our arms against any predator that might
wish to embark on conquest. But the trade of the seas must primarily be carried in great ships moving slowly to handle the
bulk and weight that cannot be carried at tolerable cost by air
or under the seas. If one nation can cut the world apart at will
by the use of submarines, it can name the tune, and all our vision
of a world united across the seas is in jeopardy.

Twice in recent history aggressors have sought to cut the world
apart by submarine and thus to prevail. Twice they have failed.
We must make sure that a third similar attempt would also fail.

For this assurance we need a vigorous program in the hands
of a virile and generously supported Navy, in the closest of
co-operation with science and technology throughout the country. We need a Navy intent on the full accomplishment of its
main mission, and not diverted by the sirens of more spectacular fields, or arguing on the defensive in regard to its importance
as compared with any other service. We need a government
organization that works for its intended purpose, a prosperous
and progressive industry, and a dynamic economy for the long
pull. Only thus can we solve the problems. Only thus can we
make it evident to the potential aggressor that we undoubtedly
can solve the problem of the submarine—so convincingly evident
that we shall not have to be put to the test.

CHAPTER VIII

THE GUIDED MISSILE

"Could not explosives even of the existing type be guided automatically in flying machines by wireless or other rays, without
a human pilot, in ceaseless procession upon a hostile city,
arsenal, camp, or dockyard?"   —WINSTON CHURCHILL
Thoughts and Adventures. 1925

EVER SINCE men fought they have thrown missiles of some sort
at one another. Yet the whole story, from the sling to the rifled
gun, is merely one of obtaining greater range, more destructive
warheads, and higher precision of aiming. For that precision,
reliance has always been on starting the missile accurately on
its flight and leaving it to pursue a free trajectory thereafter.
Guiding the missile after launching is a recent development.

It is not a recent idea, as hundreds of patents attest. But the
necessary reliability of performance of sensitive complex mechanisms was not attained until recently; it is the combination of
delicate response to stimuli with ruggedness in use that rendered
guidance possible.

There is confusion in many minds on this point, for it is easy
to believe that progress in technical fields usually waits on invention. In the field of complex mechanics, inventions are a dime
a dozen. The question of whether a device will come into being
depends upon three things: first, whether there is a practical
use for it that warrants its development and manufacturing
costs; second, whether the laws of physics applying to the elements available for use in its design allow the attainment of the
needed ranges, sensitivities, or the like; and third, whether the
pertinent art of manufacture has advanced sufficiently to allow
a useful embodiment to be built successfully. The great advance
in the evolution of guided missiles during the recent war occurred because there were new elements of precision available,

71

such as the thermionic tube, which introduced great sensitivity and versatility into design, and because we had learned to manufacture these cheaply and reliably. The opportunity for use had long existed. There were hundreds of men who could perform the necessary invention, scientists who could work out the possibilities and limitations of combinations of elements in the light of the physical laws that applied, and designers who could combine all the elements into rugged reliable units adapted to mass-production methods. These men were the true masters of the new craft. There was no field in which science and engineering combined to greater advantage than in that of guided missiles.

It will be well to illustrate what is meant by all this, for one can hardly look ahead with assurance without a grasp of what is involved in producing new devices. Suppose someone arbitrarily proposes to build a gadget having the following properties: it is to be mounted on an automobile and made to steer the car so that it will follow a white line on the roadway, to slow the car when the line turns blue, and perhaps to perform other bizarre functions. No such device will actually be built, for there is obviously no application that warrants it. But if the task appears to inventors to be interesting, dozens of ways of accomplishing it will be devised at once. Any really competent young physicist or engineer, if asked to do so, will readily sit down and put together on paper a combination that will work for the intended purpose. These combinations will be patentable, for the patent law does not concern itself much with practicability. But to build a device in a really practicable form is a very different matter. This will require the services of highly competent engineers to select the best combination and beat it into form, a succession of models and tests, and finally engineering for production. The undertaking would cost a million dollars, and the result would not be worth it. But it would be a straightforward affair from a technical standpoint and could be done.

On the other hand, suppose we have a magnetic device for detecting the presence of submerged submarines from the air, as we have, and that it is proposed to build a similar device to detect and count, from an elevation of five thousand feet, tanks concealed in a wood. This is the sort of thing that is likely to appear in "military requirements" drawn by those who neither understand what they are dealing with nor take the trouble to find out. A competent scientist will readily show that the conditions imposed are impossible. By this he means that the thing to be detected would be concealed in a chaotic variation from which it could be extricated only by fantastic means. It is necessary to use care in the use of the word impossible, in any developing art, for new means and new elements are continually appearing. But the notion that all things are possible to the scientist is amazing, and it produces foolish statements.

Of course, where there exists a real need, and also a practical possibility of accomplishment, one thing more is necessary before there will be real progress. Competent scientists and engineers must be enabled to get at it, without constraint by those who profess to know all the answers in advance. There is nothing more deadly than control of the activities of scientists and engineers by men who do not really understand, but think that they do or that they must at least give others that impression, and the worst control of all is by individuals who have long been immersed in a particular subject and have made it static.

An illustration of what is involved here is given by the history of the marine torpedo. It had a great burst of development in its early days, and by the time of the first war it had become a remarkable device. Very large power for a short run was packed into a small space eighteen or twenty-one inches in diameter and a few feet long. The torpedo was rigged with vanes that automatically kept it erect during its run in spite of the torque of its propeller. A sensitive hydrostatic element controlled elevation to cause it to run at a constant and preassigned depth. A gyroscope could be set in advance so that, after the torpedo was launched in any direction, it would take up a desired course and maintain it accurately. With all this complexity the torpedo had been shaken down so that it had become a reliable affair, safe to use, and dependable in functioning, and all this had been accomplished a generation ago.

There was a fairly obvious need to add to all the gadgetry one more element, to cause a torpedo to steer into its target at the end of its run—in other words, to make a guided missile of the torpedo. All the essential elements for doing this have long been present, and in fact during the first war there was great advance along the very lines that could quite evidently lead to it. Yet in the interval of peace between the wars, the torpedo evolved only along conventional lines, with more powerful motors of the·old type and various detailed refinements of the accepted mechanism. We do not need to be apologetic about it from a national standpoint, for the same thing occurred in every country that built torpedoes. Why was this? In the first place, the senior officers of military services everywhere did not have a ghost of an idea concerning the effects of science in the evolution of techniques and weapons, nor did they have real partnership with those who did, and it was the senior officers who wrote the requirements and controlled the whole affair. Second, the development of torpedoes had come into the hands of specialists who "knew all about torpedoes" and who admitted no outlanders to their inner councils, who had the NIH sign, meaning "not invented here," prominently in sight, and who constituted a self-contained and self-perpetuating oligarchy with its nose down on a single-track line of progress with no switches and no by-passes. It can happen in the best of organizations, civilian or military, when they grow large and old.

When the war came the old lines of resistance finally broke, and new and different torpedoes of various sorts appeared, developed by our enemies, our allies, and ourselves. Some of these were among the first of the guided missiles, and we may see many varieties in the future, now that the field has opened up.

We are more interested here in the guided missile in the air. This, too, has a long history, including obvious opportunities that were not grasped as they occurred. We have just been a little rough with those who sail the seas; perhaps we should restore the balance by also being a little rough with those who fly the air, for there seems to be something of a contest under way between these groups, and it would not do to appear partial. Progress in technology is no mere matter of youth or exuberance; it takes more than that to make it go. In fact, undiluted enthusiasm sometimes proposes the impossible and produces the monstrosity.

Way back in the early days of aircraft all controls were manual and the pilot had to fly the plane at every instant by responding to its movements and guiding it by his controls in three dimensions—the problem of the automobile driver on the road complicated by the necessity of steering in a vertical plane as well, and with no solid ground to lean on. Flying at all was an art, as of course under many conditions it still is. But soon after large planes came into use there appeared also automatic devices for handling them, notably the automatic pilot. This takes over on a long flight, steers a straight course at constant elevation, allows for all the bumps and deviations, and gives the pilot relief. It is a complex but reliable device, having a gyroscope to give it a sense of direction and pneumatic and hydraulic gears for operating the controls accordingly. It evolved from the gyroscopic compass long used to steer ships at sea.

Now here were all the elements of one sort of guided missile—an aircraft with no crew, adapted to be preset and to fly to a designated target, carrying a load of explosive. It was simple to cause such a device to drop its bomb, or itself, after a flight of specified length. It was also evident that other controls could be added; for example, to cause the craft to weave or maneuver in flight to avoid flak. Here was also a true intercontinental guided missile, for airplanes that are not coming back can readily be given long range, coming in over thousands of miles to destroy cities—the sort of thing that inspires terror in those who only partially understand. All the elements were present for such a development long before the war opened.

What happened? Practically nothing, in any country during

76 MODERN ARMS AND FREE MEN

the peace. Hitler alone had a program of development under way, from 1935 on, at Peenemünde and elsewhere, and this gave him a head start of some significance in the rocket and guided-missile fields. But no weapons in the form of pilotless aircraft appeared until the war had been under way for some years. There is significance, for our purposes, both in the things that were done and in those that were not done.

One thing that was done as the war began was to build semi-conventional airplanes in this manner, usually with all the engines and gear, and at very considerable cost, to use as missiles. They never came into use. It was fairly evident at the outset that they would not. It requires a valuable target indeed, and strong presumption that it will be hit and destroyed by a missile, to throw away an entire expensive conventional aircraft at every shot. We do not need to go into this fiasco in detail. It is an illustration of what can happen when military requirements are written by enthusiasts of little grasp, and when developments are mapped without exploration of the possibilities and limitations, most decidedly including the question of costs, by unprejudiced and competent independent technical judgment.

The single thing that did come into use in this field is of more interest. That was the V-1, the buzz bomb that began in June, 1944, to harass London. This was just a self-controlled aircraft of the sort we have described. But it had been redesigned and shaken down for its intended purpose, and that specific purpose was to bomb the city of London, thirty miles in diameter, from a range of about two hundred miles. The air frame of the V-1 was made simple and inexpensive for quantity production. Its motor was an exceedingly simple affair, cheap to build and capable of running only for an hour or two before it was worn out. If there were plenty of space, it would be interesting to pause and consider this motor, for it was ingenious and cheap, and we missed it, although, of course, its use is only for short-duration purposes. The controls of the missile were exceedingly simple, for no great accuracy was necessary to hit a target the size of London from that distance. The Germans intended to launch these devices at the rate of about three thousand per day, and to keep it up for weeks, against London and against the Channel ports, and they intended to do so before we could launch our invasion of the Continent. Had they been permitted to do so, they might well have stopped the invasion. The missiles carried about a thousand pounds of high explosive apiece. There was no guarantee that the warhead would not contain toxic materials instead. It was a real threat.

It was a much more important weapon to the Germans than it would have been to us, for they had the target. In fact, to us, it probably would at no time have been worth its cost, which was not inconsiderable when all factors such as handling and launching sites were included. We had no situation in which highly important targets could be reached only by such a missile with very low precision, for we could reach our targets much more effectively by conventional bombardment methods. But the fact remains that though in the interval of peace or in the early days of the war we had no assurance that we should not need such a weapon, we did not develop a missile like the V-1; moreover, when the situation became clearly one in which the device had little valuable application, we did develop it, though we did not use it. We had a blind spot on this whole affair, and it will do us little harm to admit it and still less harm if military and civilians join frankly in the admission.

What was not done by the Germans is also of interest, after they are given full credit for what they did do. Their program was reduced to relatively small proportions and delayed for nearly a year by bombing of their laboratories and manufacturing plants and especially by bombing of their overelaborate launching sites on the Pas de Calais. They finally developed simple mobile launching rigs, but learned to do so the hard way. When at last their attack was launched, the invasion was well under way, and the bombardment they achieved was on a shoestring basis compared to what they had intended. Incidentally, in passing, we owe much credit to those scientists and military men of our intelligence groups who traced the German plans and progress with some accuracy throughout. One amusing aspect of this effort was that the variation in volume of Nazi boasts in neutral publications concerning their terror weapon furnished a fair index of the success of our bombing in interrupting progress. The greatest error of the Nazis, however, was that the device they finally launched was highly vulnerable in the light of concurrent developments in antiaircraft artillery.

Even so, we had a regrettable difficulty. The batteries of antiaircraft artillery that were set up on the coast of England to meet the buzz-bomb threat were necessarily diverted to Normandy when the invasion occurred, and London, left wide open for a while, suffered heavily until they could be replaced. When they were, with radar-controlled guns, electrical computers, and proximity fuzes, the buzz bombs as then used made almost ideal targets, flying slowly at constant altitude and in a straight line. When the defense dispositions reached their climax they brought down some ninety-five per cent of the buzz bombs that came within range, and they repeated or bettered this performance later at Antwerp. The buzz bomb was very decidedly countered. If the Germans had correctly estimated the nature of the defense at the time of the introduction of their new weapon, they would at least have built it so that it would weave in flight in the presence of antiaircraft fire. More soundly, they would not have diverted effort from more important things to build it at all.

They were, however, much nearer to reality in the matter than we were. Part of their lead was owing to the fact that they had an almost ideal target for such a missile staring them in the face (even if the importance of that target was exaggerated by Hitler's hate), and we did not. But most of the difference of performance on this particular weapon was owing to the fact that the Germans estimated costs at the outset and built a comparatively cheap device; whereas when we came to the subject at all we did not balance costs against probable effect. Their final failure with the buzz bomb occurred because, shrewd as they were regarding costs, they failed to estimate the trends of the entire related art.

There is a common notion that during war costs do not count. There is no greater fallacy. The error comes from the belief that civilian resources are unlimited. They are not. Costs are more important in war than at any other time, for the need for over-all effectiveness is then more imperative. Every development and procurement of a useless device detracts just that much from progress along highly necessary lines. If we are going to fire a missile at the enemy, we should be very sure before we devote a large amount of manpower and materials to it that it is going to harm the enemy more than it harms us, that the damage it causes will interfere with his ability to continue the struggle more than it costs us from the same point of view. No industrial concern would launch a development without prior cost studies or without causing it to run the gantlet of tough-faced engineers with sharp pencils. Military organizations can proceed without this unpleasant preliminary skirmishing, and often do. When this happens performance is bad. It is equally fallacious to under-take a development without estimating its performance in the light of conditions that will apply when it actually gets into use, allowing for development in other lines in the meantime. We see all this well illustrated in the case of the buzz bomb, and it is essential that we keep it in mind as we examine the possible guided missiles of the future.

There is one more point to consider in regard to costs, for it is easy to be led astray. In times of peace the reasoning of the military services resembles somewhat that of industry; dollar costs are paramount, for dollars are limited. There is a difference, because the military have to estimate values in terms of a vague possible future use, so that their comparisons are between one weapon and another, relative rather than absolute; whereas industry must estimate a public demand for an explicit article. When war comes, a rapid transition occurs in the basis of estimate, and the military estimate is hard to make. Dollars are then likely to be, at first at least, practically unlimited. The dollar loses its concrete significance. The problem is to get on with the war at maximum speed and win it with a minimum number of casualties and a minimum of damage to property and to the economy generally. Only a limited amount of materials and manpower is available, and the whole problem is to distribute these in optimum manner to secure the desired result. This requires a balance between numbers of men in civilian life, at the bench, or in the laboratory and the numbers in uniform in the field. It requires balance between new development of weapons and production for immediate use. The dollar is a convenient measure, but not a limit. Detailed balance and comparison can be carried out by the military services, bolstered by recruitment of managers and other experts from civilian life. But the broader judgments call for an active, able war cabinet. We have never had such a war cabinet in fighting our wars. It would be an exceedingly valuable asset, should we have to fight again.

In discussion of the future of guided missiles, one more principle should be examined briefly. There are certain operations that can be performed better by a machine than by a man, and by the same token there are certain things in which man can outperform the machine. No small part of the skill involved in mapping out developments, where complex devices are involved, depends upon a keen appreciation of this matter. A machine can go where a man cannot. A machine does not have to come back. A machine can be depended upon to respond to conditions by explicit actions in simple cases more reliably than a man would respond. A machine can even be supplied with a rudimentary memory and enabled to exercise judgment of a sort in simple cases. But a machine of reasonable complexity cannot be expected to exercise judgment under highly involved conditions or conditions other than those for which it was designed.

Thus, a pilotless airplane could be made to fly smoothly thousands of miles in spite of buffeting and weather vicissitudes. It could be made to take off and land automatically. It could even be made to carry out its own navigation to a distant target, for example by using loran, and to drop its bombs there. By resort to extremes of mechanization it could even be made to fire its

guns or other missiles at approaching enemy aircraft. All this
without a crew aboard. All this could be done, and other features
could be added, all at enormous expense for development and
very large cost per unit. But it would not be worth doing. Compared to an airplane with a well-trained and skillful crew aboard,
such a craft would be an easy victim to enemy defenses. Like
the Japanese pilots, rigidly trained to meet an explicitly defined
set of conditions, it could perform well under those conditions,
but it could not adapt to meet a change in them.

There is in some minds at the present time an increasing fear
of intercontinental guided missiles. The device just described is
not only such a missile; it is the only type of such a missile that
can now be built. We may need to fear intercontinental bombing
by manned aircraft at high altitudes—just how much will depend
on many factors, and we shall examine some of them later. But
there need be little fear of the intercontinental missile in the
form of a pilotless aircraft, for it is not so effective from the
standpoint of cost or performance as the airplane with a crew
aboard. And when we come to the intercontinental missile of
the other type, the missile that flies in an enormous trajectory
high above the earth, we shall see that our fear of that also may
be reasonably tempered.

Two developments have led to the guided missile. One of
these is the matter of controls, with which the discussion thus
far has been mainly concerned. The other is the appearance of
new ways of projecting a body through space. One of these is
the rocket, now to be considered.

When a missile is fired out of a gun the entire impulse is given
during the brief fraction of a second that it is in the barrel. Pressures and accelerations are enormous. The force on the missile
may be some thirty thousand times as great as its weight. It gets
under way in a hurry. To produce these results involves both
large weight and cost, and the guns themselves are highly refined pieces of precise construction. A rocket, on the other hand,
carries its own propelling charge, and this burns relatively gradually and produces the propulsive force over a comparatively

long interval by the reaction of gases escaping through a nozzle to the rear. Pressures are relatively low, construction need not be highly precise, and the device is cheap to build. Moreover, no gun is needed, merely a pair of simple rails or a tube to start the affair in the desired direction.

The simple rocket, such as described, has not attained the velocities of shells from guns or the precision. Yet it had many applications, where a heavy volume of fire was needed at relatively low range, where high precision was not necessary, where the weight of a gun was undesirable or prohibitive. It was developed for many purposes, with the faint support at first of those who know all about guns. The Russians did a great deal with it for field-artillery purposes. We mounted it on all sorts of vehicles, and it was especially useful for beach bombardment in amphibious operations. Its greatest success came when it was mounted on aircraft, for it gave them the hitting power of large guns without the guns. For attack on tanks or surfaced submarines it made the airplane a formidable instrument indeed.

These were all short-range missiles, however. It hardly paid from a cost standpoint to give them guidance to improve their otherwise mediocre precision, for they were to be used in great numbers and they had to be inexpensive. It did pay decidedly to supply them with proximity fuzes, and we did not reap all of such benefit that was available. A rocket thus equipped is a deadly affair when fired from an interceptor plane against a bomber. The bomber is not in nearly so advantageous a position to use it for defense, for aiming a rocket is difficult and is best done by aiming the entire airplane. A rocket, moreover, is not likely to behave well when fired across the high-velocity air stream as it generally must be when used defensively from a bomber to catch the attacking fighter coming in abeam. In such a case, also, the fighter has the advantage of its own speed, which adds to the speed of the rocket to give a high resultant velocity, whereas the bomber does not. Rockets can be made larger than those for air fighting, to fly at very high velocities, and they can thus be given ranges much greater than that of a gun, though at considerable expense and loss of simplicity and ruggedness.

The important example of the large high-velocity rocket device was the German V-2. Unlike the V-1, this was a true rocket, not a pilotless aircraft. It was an expensive and complex affair, involving great ingenuity and remarkable engineering skill. Its development went back to about 1935, and it appeared during 1944. It was intended primarily for bombardment of London from a range of about two hundred miles. Like all true rockets it carried its own fuel and oxygen supply, in this case in the form of liquids, pumped into a combustion chamber by pumps that developed thousands of horsepower for a short interval. Its propulsive force was produced by a great blast of hot gases ejected to the rear. It flew in a great arc, seventy-five miles up into the stratosphere, and descended at three thousand miles an hour, well above the speed of sound, so that its arrival was not heralded, and one heard the roar of its passage through the atmosphere only after its ton of high explosive had detonated. It was a terror weapon.

It was a guided missile in a certain limited sense. All of its propulsion occurred in its first flight up through the atmosphere, and this was somewhat erratic, so that its flight was watched by radar during this first brief interval, and radio signals were transmitted to cut off its propulsion at the optimum instant. After that it flew in a free trajectory, and its precision was far from good. It could usually hit London at that range, but it most decidedly could not hit any assigned target such as a factory or an airfield.

Did it pay the Nazis to develop and use the V-2? The matter can be argued at length. From a strict damage-and-cost standpoint the answer is no. If the V-2 had destroyed a square mile of London, it would have done so at an overall cost to the Germans, in terms of interruption of war effort, far greater than the cost of rebuilding to the British. The ton of explosive a V-2 delivered was negligible in comparison with the ten-thousand-ton aircraft raids of that time. Development and production of the V-2 called for the very skills, facilities, and materials that could have been used to much greater advantage in the program of jet pursuit aircraft, which, if thus used, could have been embarrassing indeed to our progress in bringing Germany to her knees. On the other hand, the V-2 certainly produced terror, coming out of the skies at any hour, without warning, and without the possibility of interception once it was launched. We know now, although we did not know it at the outset of the war, that the spirit of a resolute people is not easily broken by all the terrors of conventional mass bombardment, for we have the record of the magnificent performance of the people of England in that regard, and of the people of Germany or Japan, for that matter. We also know that we originally far overestimated the importance of the physical damage caused and underestimated the rapidity with which it could be repaired. But the high-flying missile introduced another element, a psychological one, into the picture. It came late, when nerves were frayed. It by no means produced panic, but it did add to stress. From that standpoint it may possibly have paid out, though even this is highly doubtful.

The Germans erected huge launching sites, at great cost and effort, for the handling of V-2 missiles in quantity, but they did not use them, for by the time they got to it the sites were not usable. Instead, the Germans resorted to mobile launching arrangements, which could move out by truck to a road intersection or the like, launch their missiles, and leave promptly. The flight of the missile could often be seen on a radar screen as soon as it started. The British therefore had aircraft ready in the air and relayed information to them rapidly, so that they could attack the launching spot. This was sometimes successful. But in general there is no defense against the high-flying missile except to prevent its being launched, any more than there is a defense against the shell after it has left the gun. Means of interception in flight are not inconceivable, but they are certainly very difficult.

We are hence decidedly interested in the question of whether there are soon to be high-trajectory guided missiles of this sort spanning thousands of miles and precisely hitting chosen targets. The question is particularly pertinent because some eminent military men, exhilarated perhaps by a short immersion in matters scientific, have publicly asserted that there are. We have been regaled by scary articles, complete with maps and diagrams, implying that soon we are thus all to be exterminated, or that we are to employ these devilish devices to exterminate someone else. We even have the exposition of missiles fired so fast that they leave the earth and proceed about it indefinitely as satellites, like the moon, for some vaguely specified military purposes. All sorts of prognostications of doom have been pulled from the Pandora's box of science, often by those whose scientific qualifications are a bit limited, and often in such vague and general terms that they are hard to fasten upon. These have had influence on the resolution and steadiness with which we face a hard future, and they have done much harm, vague as they are. But this one is explicit, and we can treat it.

The German V-2, of two-hundred-mile range, carrying a ton of explosive, was nearly at the limit of effective range for a chemically propelled single-stage rocket. Its range can be substantially increased in three ways. First, its payload may be decreased until at four hundred miles or thereabouts it will carry no explosive whatever; second, its size and its cost may be increased to carry the present payload a greater distance, though range does not go up in direct ratio to size; third, it can be made into a multistage affair, with an enormous rocket giving birth, when it has done its work, to a smaller one that proceeds from there, and so on. In other words, great range and payload can be attained at very great cost. If we are content to pay millions of dollars for a single shot at a distant target, it can be done in this way for any stated distance.

No escape from this dilemma is to be had in new chemical fuels. We are already near the limit of the amount of energy that can be chemically packed into a given weight. But I can immediately hear the answer that this situation is all changed now that we have atomic energy available. Perhaps the best reply to this comment is that it is admitted to be a close thing whether atomic energy can compete on a cost basis with chemical fuels for producing energy for commercial purposes under the relatively easy conditions on the ground, where weight and space are not limited, and it is not necessary to release all the available energy in a brief interval; and we have already stated that a rocket can be made to fly far if one disregards costs. The missile that flew a thousand miles high above the earth might burn itself up like a meteor when it again hit the atmosphere on its descent, but it can be built.

But can such a missile be made to hit anything at the end of its flight? The V-2 could be made to hit with reasonable frequency within fifteen miles of a point of aiming at a range of two hundred miles. A similar missile flying two thousand miles could be depended upon to hit within 150 miles of its target with reasonable frequency. This probability can certainly be improved upon. Conceivably, the missile could be given instruments of precision and could then operate automatically for navigation and guidance. It might do nearly as well as a manned aircraft flying above the clouds, dependent upon the stars or upon signals from its home territory thousands of miles away, and with no guidance, but possibly confusion, from the areas near its target. Perhaps it could add sights and homing aids, as the aircraft might. It could then hit a target, perhaps within ten miles, perhaps even within a mile or two if all went very well indeed.

Its cost would be astronomical. As a means of carrying high explosive, or any toxic substitute therefore, it is a fantastic proposal. It would never stand the test of cost analysis. If we employed it in quantity, we would be economically exhausted long before the enemy. It makes little difference in war whether one's people, facilities, and materials are destroyed or whether they are employed in making devices that are then destroyed without advantage.

But it will carry atomic bombs! Here, if at all, the intercontinental rocket might conceivably enter in time. But as long as atomic bombs are scarce, and highly expensive in terms of destruction accomplished per dollar disbursed, one does not trust them to a highly complex and possibly erratic carrier of inherently low precision, for lack of precision decidedly increases such costs, as the following chapter will indicate.

For the near future, the really important and significant field of guided missiles lies in much shorter ranges, above those readily handled by guns, but not so large as to run up size and costs to prohibitive heights. In this range they have applications to extend the range of field artillery, but they have much more important applications to air warfare.

A number of new engines for propelling aircraft were developed during the war, the jet and the turbojet and finally the ram-jet or athodyd. All these used the principle that is at the basis of the gas turbine, namely that of compressing a gas, heating it, and then deriving from its expansion more energy than was needed for the compression. The basic idea is older than Ericsson, who in fact handled it rather ineffectively, but its fruition into practical form necessarily waited until the metallurgists had produced materials that would stand the gaff. The simplest of all these devices by far is the ram-jet: the stovepipe with a fire in it. It is simply a properly shaped open tube. Its rapid motion through the air produces the necessary compression at the front end, without any other mechanism. In the middle is a fire to heat the gas, and there is some difficulty in keeping it burning. The rear of the tube is a nozzle where the expanding gases acquire velocity and produce the push. A relatively small unit of this sort, about the ultimate in simplicity, can produce thousands of horsepower—once it has reached a speed high enough to secure the needed compression. The trouble is that this speed has to be very great; some three thousand miles an hour is needed to make the ram-jet work really well. Hence it must be shot out of a gun, or driven in the first of its flight by auxiliary rockets or the like, and this fact entails complexity.

This fact, moreover, makes the ram-jet of no use for propelling

ordinary aircraft, for even the most powerful of these comes far
short of three thousand miles an hour. Nor is the ram-jet applicable to the high-trajectory, long-range missile, for it needs air
for performance, and there is very little air indeed at the heights
at which such missiles fly, although the ram-jet can obtain air
for its needs at any altitude at which an aircraft can do likewise.
Lastly, the fuel economy of the ram-jet is none too good in the
best of circumstances.

To drive a relatively short-range, very high-speed missile,
however, the ram-jet is almost ideal. This sort of device can be
shot out of a gun or driven by rockets to reach the desired high
speed at which the ram-jet works best, and thereafter it can
travel at bullet speed or higher and keep going for ten or perhaps a hundred miles. Can it hit anything? There is a vast difference here from the problem of the long-range missile. Over
short ranges, at least, it certainly can be aimed in the vicinity
of the target, and from there it appears that reasonably complex
homing devices might take over and bring it to the point where
its proximity fuze could detonate it with lethal effect to an
aircraft. The device is in its infancy, and only time will tell what
can be accomplished with it in a practical way. But, as the world
has been told, such devices have flown.

Its significance lies in the fact that it appears to be the great
future enemy of the bomber, and it may in time turn the tide
that began to flow at Kitty Hawk in 1903 and that reached its
crest in the bomber fleets that darkened the skies of the last war.

Let us assume that devices of this sort can indeed attain a
range of, say, fifty miles, and that with that range they can
home with precision and a reasonable percentage of hits upon
a bomber. Certainly the bomber cannot maneuver out of the way
of a missile that comes in with the speed of a shell from a gun.
Certainly it cannot survive the explosion of a hundred pounds
or so of high explosive detonated in close proximity. Perhaps the
bomber can jam the controls of the missile, if it knows the missile is coming and knows its combinations of frequency and the
like, but the advantage here seems to lie with those on the

ground, who can choose a variety of systems at will, so that if one is countered others are not. Perhaps the bomber can in some manner avoid going near the installation on the ground, but if it tries such evasiveness there appears to be no reason why the appropriate missiles cannot be carried against it by jet planes promptly vectored to a threatened point upon radar warning. Perhaps the bomber can slip in on the deck, close above the trees, to avoid defensive radar and guided missiles. Perhaps it can, but the sneak raider close to the ground is a far cry from great high-flying fleets, and if we can counter these we ought ultimately to have means for taking care also of the raider that flits about just over the trees.

The days of mass bombing may be approaching their end. If so, it is a good thing for the world.

CHAPTER IX

THE ATOMIC BOMB

"... neither the atomic weapon nor any other form of power and force constitutes the true source of American strength ... if we embrace this escape from reality, the Myth of the Atomic Bomb, we will drift into the belief that we Americans are safe in the world, safe and secure, because we have this devastating weapon—this and nothing more. We will then tend to relax, when we need to be eternally vigilant." —DAVID LILIENTHAL *Commencement Address,* Michigan State College. June 5, 1949

THE HISTORY of the development of the atomic bomb has been well told, and discussion of the bomb itself soon reaches the boundaries of the technical matters that may not properly be analyzed publicly. The atomic bomb is such a vast and controversial affair and there has been so much discussion concerning it that a brief summary, such as is needed here, is bound to be incomplete and only in part supported by evidence or argument. Still, there are plenty of extensive treatments available, and we need touch only on salient aspects. No man can predict the future of the bomb with certainty; yet, in the present and evident disagreement between experts on the subject, it is possible to see much of common ground; and this summary, while it will be disagreed with also, will serve our purpose of an overall view of the relationships between weapons.

The principal point is that the atomic bomb is for the immediate future a very important but by no means an absolute weapon, that is, one so overpowering as to make all other methods of waging war obsolete. It will remain in that status as long as there are not great stocks of bombs in the hands of more than one power, and even if this should occur it may remain in that category unless there also then exist means for delivery of the bombs onto enemy targets with at least a moderate degree of assurance.

90

The atomic bomb was indeed a terrible weapon. As things stand at the moment it is still a terrible weapon, and one to be feared. It came by surprise as a great shock to a weary world. Its spectacular nature, and the strange toxic effects that accompany it, overwhelmed our calm reason for a time, and it is not astounding that the bomb promptly became a basis for predicting the end of the world as we know it. There has now been an interval, and we may assess the bomb more objectively.

It was indeed the bizarre nature of the bomb, and the uncanny sort of future it suggested, rather than its actual results in the war, that impressed people. The fire raids upon Japan were much more terrible, they reduced a far-greater area of the frail Japanese cities to ashes, they caused far-greater casualties among civilians—panic, the crush of mobs, and horrible death; yet they occurred almost unnoticed and created few later arguments. The moral question was hardly raised regarding the fire raids, yet that question is substantially identical in the two cases. It was fear of the future that concentrated attention on the atomic bomb.

The use of the atomic bomb ended the war. Without doubt, the war would have ended before long in any case, for Japan had been brought nearly to her knees. Her fleet had been destroyed. Her commerce had been cut off and her ships destroyed by our submarines, our mining campaign, and our attack from the air. Her planes had been nearly driven from the skies, and her cardboard cities were wide open to mass bombing. In these circumstances, we needed merely to keep the pressure on, at minimum cost in casualties, and she would most certainly have collapsed into utter impotence, whether she formally surrendered or not, and she could then have been contained by our Navy without risk until she did yield. Yet in the face of these facts we had planned and had in motion at the time the bomb fell a vast program of invasion by land forces. With the experience of Okinawa, with the demonstrated fanaticism of the Japanese and the proved ability and intent of their masters to whip them into a frenzy, it was clear that such a campaign might have cost hundreds of thousands of casualties among our troops. The ships and troops were moving, the first waves of troops had been told off for the sacrifice on the beaches, and all the juggernaut of massive land war was under way. Into this acute situation came the atomic bomb, with a maximum of surprise and dramatic effect and in such circumstances that the Japanese war lords, able as they undoubtedly were to steel their people to mass suicide, had no opportunity to prepare them for it. Two bombs went off, and the war ended. It is useless to argue how much they advanced the end. Certainly enough to save more lives than they snuffed out and more treasure than their use cost. Certainly they ended the war under such conditions that we were free to begin the rehabilitation of the Japanese people rather than forced to undertake the conquest of a starving desert inhabited by a broken lot of physical and mental wrecks. But the abrupt ending after six years left the world gasping. No wonder, then, that we found it difficult to discuss an atomic future with objectivity.

We need to examine atomic fission in two aspects: first, as merely a new way of manufacturing high explosive in concentrated packages, and, second, as a weapon possessing extraordinary attendant toxic effects.

From the first standpoint we need merely to compare the atomic bomb with ordinary high explosive, the TNT or the somewhat more powerful RDX of the last war, on the basis of cost and delivery. It has now become fairly clear that atomic bombs are expensive and will remain so. While a cost comparison must be based on many factors such as cost of delivery, industrial capabilities, and the efficiency of governmental operations, and while time may alter these, it has now become fairly clear that production of atomic weapons requires such major expenditures and such major effort that they cannot be afforded at all except by countries that are very strong economically and industrially. In this country we have striking evidence of the high original cost of plants and methods, for we paid the bill. Now, too, it has become known that, at the time of Hiroshima, and after this vast expenditure had been made, we had no great stockpile for the devastation of all of Japan's war potential, although she undoubtedly quit as suddenly as she did because of the belief that we had. In fact, the number of bombs was then startlingly few in the light of the effort that had been applied. One can never say that short cuts will not be found in manufacture that will change this situation, that radically different processes will not reduce by a large factor the plant and man-hours necessary to build a bomb. But one can say that many of the best physicists, chemists, and engineers of this country, and of Great Britain and Canada, have been seeking such a process now for nearly a decade. They have examined every conceivable approach, and they have not found it. It appears highly probable that atomic bombs are going to remain very expensive for a long time to come. It has also come to be known that the raw materials for bomb manufacture are limited, not so much in quantity as in quality. There is some concentrated ore, but not much. There is a great deal of very dilute ore, so dilute that the process of concentration and recovery is laborious and expensive. Thus much time or extremely large expenditures of money will be required to overcome limitations on raw material.

Out of this situation has come the gradual recognition that not for some years will there be a number of great stocks of atomic bombs as a part of the world's armament. It has also been grasped that the task of repeating what this country did under the pressure of war is no mean task and requires years of effort. Thus the time has been moved ahead when there may be two stocks of bombs of comparable and substantial size, and we have more breathing time than we once thought. There is a high probability that there are some years, perhaps quite a few, before the question of two prospective belligerents frowning at each other over great piles of atomic bombs can become a reality.

The time estimate depends, of course, on how fully we think our adversaries may put their backs into the effort, how much they are willing, or able, to reduce their standard of living in order to accomplish it. They lack men of special skills, plants adapted to making special products, and possibly materials. As we shall discuss later, they lack the resourcefulness of free men, and regimentation is ill adapted to unconventional efforts. On the other hand, their tight dictatorship can order effort, no matter how much it hurts. But we do not need an exact estimate; it is sufficient to note that opinion now indicates a longer time than it did just after the end of the war. The problem is not altered in its nature by this more moderate estimate; it is certainly less critical and immediate. This is important, for there are other forces active, and the conditions affecting the mass use of such bombs may well be entirely different before the issue is faced squarely. We may then, in fact, be living in a different sort of world.

The cost of trinitrotoluol, the TNT that is the most common high explosive, is less than a dollar a pound when it is manufactured in quantity. Built into bombs, delivered on a target hundreds of miles distant by an intricate aircraft manned by a highly trained crew subject to the attrition of war, its cost may well mount to hundreds of dollars a pound. Behind this ratio lies the opportunity for gross fallacies in reasoning.

This is not to say that atomic bombs, measured in terms of the destruction caused, and including in costs the costs of delivery, may not be less expensive than conventional high explosive. Rather it is to point out that between these costs there is no spread so great as to make the atomic bomb overwhelming when considered from the mere standpoint of the cost of causing destruction. Moreover, it is not to ignore the important element that the cost of atomic bombs is largely a peacetime cost, for they cannot be manufactured in a hurry during war, as can high explosive. Costs, that is, effort in terms of labor and materials, are necessarily spread over a long interval to produce atomic bombs, and this fact greatly affects our reasoning concerning them.

We have nearly forgotten that long before Nazi might was being built up for conquest, in the days prior to the last great war, a revolutionary theory of warfare was advanced by the Italian army officer Giulio Douhet. Some parts of it sound strangely familiar today. It held that all means of making war other than strategic air bombardment had become obsolete. At the outbreak of war, Douhet argued, great fleets of bombers carrying high explosive would completely devastate the enemy country. This phase might be preceded by air battles, but one contestant would then secure control of the air, and after that he would hammer the enemy into submission. Neither armies nor navies needed to move. The loser in the air battle would submit, or his entire structure of civilization would be reduced to rubble, his casualties would be enormous, and his whole organization plunged into chaos, so that he could only weakly resist invasion and conquest. It was fear of this doctrine and of the air fleets being built by the Nazis that accounted principally for the terror that seized the rest of the world in the middle 1930's and that accounted for appeasement. Unable to rally their own people to build air fleets of their own, with the United States far away and apparently indifferent, softened by the arguments of those who felt that any sort of submission was preferable to war of that sort, nations were drawn into the fiascoes of Abyssinia, the Rhineland, and Munich and the inevitability of war. It was terror of the air fleets that weakened the will to resist.

The actual outcome is history. A few men of vision, in spite of the currents of public opinion and apathy, saw to it that Britain had radar and advanced pursuit ships. The blow fell, Britain was sorely damaged, the bombers were turned back, and that wave of barbarian conquest was halted.

But the later stages are most important from our present standpoint. Britain and the United States built enormous fleets of bombers, extending their industrial competence to the utmost to do so. They gained air dominion over Germany, and they pounded the cities and transportation systems of Germany as no other cities have ever been pounded. Still it was the great land armies in a magnificent campaign that defeated Germany. Air attack became a part of this effort, an essential element without which the crossing of the Channel and the fight to the Elbe could not have been accomplished in years if at all, but nevertheless a part of a whole. Since then we have had a survey of bomb damage and the reports by the d'Olier committee, and we can appraise the situation objectively. Not all of us can, for the enthusiasts of air power are often not stopped by such a minor obstacle as an obstinate fact, but groups of analytically minded military men and specially experienced and trained civilians can view the entire record and come to clear conclusions if they will. Even with those facts which are on the public record it is not difficult to draw reasonable, if tentative, conclusions.

These conclusions are that, as the techniques of bombing then stood, the mass bombardment of enemy cities, in the face of determined defense, as the sole means of bringing victory over a foe of equal or comparable strength was a delusion, and such general bombing was not worth the extreme cost and effort it entailed. On the other hand, selective bombing of key facilities, of transportation, fuels, and critical manufacture, at a time when control of the enemy air had been secured, very decidedly paid dividends. In looking at this matter we should not draw unsound conclusions from the Japanese case. When our full strength was brought to bear, Japan was by no means an equal adversary. But the lessons of the bombardment of Germany are clear. As matters then stood, and as between approximately equal antagonists, mass bombing was of dubious value, and selective bombing was an essential element of co-ordinated attack.

From the standpoint of its explosive effect alone, substitution of the atomic bomb for ordinary high explosive alters this situation in three ways. First, it may decrease the overall total cost of producing a given amount of destruction in bombing great cities. Second, it brings about a given amount of destruction instantaneously rather than over a longer interval. Third, because of the large proportion of the cost of atomic bombs involved in the production of the fissionable material itself, there is an accumulative factor of importance. This last point warrants close attention. Money spent in peacetime for weapons that deteriorate or become obsolescent may be money largely lost. But fissionable materials deteriorate very slowly, and they can be stored. In preparing for a war that may be remote, there is thus a considerable reason for concentrating on the weapon that is sure to last until needed, and this is especially important for democracies that fight when they must, not when they choose.

Does a mere reduction in the cost of the explosive itself tend to place us again, some time in the near future, in the position where another Douhet can frighten the world by the specter of area bombing by fleets of aircraft, where another Hitler can again build up his power by seizing neighbors while the world stands by in fear, until he is strong enough to undertake armed conquest? This depends upon many matters, but it depends strongly upon what happens to the cost of delivery of explosives from the air, for the full cost of placing bombs on a target includes the costs of carrying them there as well as of the explosive itself, and costs in this case connote the extent of the national effort in terms of men, materials, and manufacture necessary to carry through such an effort to a finish. If some one nation, at the expenditure of a small fraction of its total national effort, can build the air fleets and the bombs necessary for the purpose, if these air fleets can get through to their targets with such facility that a considerable fraction of their loads can be accurately delivered, if this fraction is indeed capable of substantially destroying the warmaking potential of a neighbor within a short interval, beating down defense and retaliation, then there is a situation where the existence of the atomic bomb may indeed be the determining factor in the course of events. If an aggressor nation alone were in such a position the threat would be imminent and severe. But if the cost of these preparations is beyond the resources of an aggressor nation, so that it would exhaust itself and collapse internally during the build-up process, if conceivably defense should build up in potentiality at a rate comparable to that of attack, and if the true nature and extent of the threat are accurately estimated by the world, this method of conquest will be ended. If the cost of such an effort were merely that of the explosive itself, the advent of the A-bomb might have been determining. But the greater part of the problem may well lie in the means of delivery, and this depends upon the status of defense, so that the question is by no means one to be answered offhand. As we look beyond the near future, we need to consider both aspects of the matter.

Some of the possibilities are indicated by the trends that existed toward the end of our campaign in Europe. Toward the end of the war very significant changes began to enter into the situation. The Germans began to get very fast jet pursuit ships into the air. Whenever they appeared, we began to take losses in our bomber force. The simple fact was that the jet pursuits came in so fast that the armament of the bombers was of little use against them. Moreover, accompanying fighters could hardly move into defensive position before the attack was pressed home and the jet pursuit ship was gone. It is well to examine this development with some care, for the trends profoundly affect the way in which we view bombing in a possible war in the future.

The Nazis by no means made the most of the possibilities of the jet pursuit plane. In the first place, they built only a few of them, partly because their production was seriously disrupted by that time, and because through Hitler's hunch and hate they were off on a wild hunt on V-bombs, which used the same sort of production effort. Next, they did not have good radar, and this is the core of adequate use of very fast interceptors, for such craft need ground radar to bring them near the bomber and airborne radar for the final approach, especially at night. Even in the daytime the eye is not much of an instrument in a ship making four or five hundred miles an hour, which must begin its run-in on a bomber from a distance at which the bomber is barely visible to the naked eye. The Nazis took their young technical men off radar, in the early days when they felt sure of a short war, and by the time they replaced them it was too late, so that they lagged in this field throughout the war.

In passing, it may be noted that we came perilously close to doing the same thing, for our manpower control was in the hands of those who thought in terms of masses of men and equipment rather than in terms of awakened invention and development. We managed to maintain our staffs during the war in spite of the system and not because of it, and it consumed far more time and energy than it should have. A group of enlightened military men with the strong support of Secretary Stimson did produce some order out of the chaos. But the problem of keeping young scientists in the laboratories was one of the toughest and most irritating problems we faced in the war. The British learned their lesson in this important matter in 1914-18. It is not fully clear that we have learned it yet.

On another phase of air defense the Nazis also lagged very seriously. They never built a proximity fuze for antiaircraft shell or for their antiair rockets, which they used devastatingly against us at Schweinfurt. The techniques of this fuze were just too much for their regimented science, and manufacture of a device as intricate as this was hardly a proper task for slave labor. Toward the end of the war they had a number of very active projects on antiaircraft guided missiles, but none of these came into use. In other words, they quite completely bungled the modern phases of air defense. To anyone who knows the power of these defense instruments the answer is clear. If they had given these things the attention they deserved before it was too late, and had used the skill and resourcefulness that we rightly expected of the German nation in the days before it came under the Nazi heel, they could have stopped our bombing. They would have stopped it. We could not have stood the attrition they could have caused, from either a morale or a material standpoint, and we would have fought the war through to a finish, with air power limited to the support of the army and navy in their customary missions, and severely limited even in this regard.

But, it may be claimed, all this would have been changed if we had developed the jet plane to its full advantage, for it was really the jet plane that tended to throw the advantage to the

German defense. It is said that Great Britain and the United States lagged in the development of jet planes, and that if we had countered jets with jets the situation would have been different, for we could then have protected our bomber fleets. Lag we did. Part of this lag was probably owing to the fact that the development in this country was in the hands of those who knew too much about the subject; one seldom gets unconventional advance in full measure except by the introduction of new blood. But most of it was owing to the fact that, at that stage of the war, we did not acutely need jets and therefore did not bring weight to bear on their development. A jet engine was, and still is, a heavy consumer of fuel; its advantage is that it gets a lot of power into a small package for a brief time. It was not adapted to driving long-range bombers, where the excess fuel load subtracts from the payload—subtracts so much, in fact, that it then completely offsets the saving in engine weight—and for the same reason it was not adapted to long-range fighters to accompany and protect bomber fleets. This condition has changed somewhat since, thanks to better fuel economy of turbojets, but it was then distinctly the case. The jet was adaptable to the plane of short duration and very high speed, namely the defensive interceptor of bombers, and it was not adaptable nearly so well to the accompanying fighter, which had to fly distances comparable to those of the bomber. There is a fundamental factor here that favors the defense, namely that the short-duration defensive aircraft must be expected to outperform the fighter that has the burden of a long accompanying flight.

There is more to it even than this. Consider a fleet of relatively slow bombers, accompanied by fighters of very high top speed, and attacked by pursuit ships of equally high speed. Remember that these very high-speed craft have a turning radius of as much as a mile or two, and that it is decidedly hard to see a plane a few miles away even under favorable conditions. Remember, too, that the defending pursuit ships have all the advantage of ground radar, in early-warning sets and sets for ground-control interception, to place them in most advantageous position, while the bombers and fighters have only air-borne radar of very limited range. It is highly doubtful, under these conditions, whether the accompanying fighters would get into the fray at all. Even if they did, they would have unequal combat, and their bombers would be sitting ducks to the interceptor that eluded them. It is a reasonable conclusion from all this argument that bombing, by bombers of the speeds and altitudes of the last war, is now obsolete against a fully prepared and alert enemy.

If it is, then we have less reason to be terrified by the thought of the A-bomb delivered by fleets of bombers. We cannot discount this method of delivery, of course, and determined attacks with great attrition could undoubtedly be pressed through against highly important targets. Moreover, the means of defense is highly expensive and it must be alert. We have to think of changes in bombing techniques that might again swing the balance in favor of the attack. We have to think of sneak raids and surprise attack. We have to think of other means of delivery than dropping bombs from the air. But the specter of great fleets of bombers, substantially immune to methods of defense, destroying great cities at will by atomic bombs is a specter only. There is a defense against the atomic bomb. It is the same sort of defense used against any other type of bomb. The advent of atomic explosives alters the prospect very substantially in one direction. There are changes under way that alter it in the other direction, in favor of the defense. We need to be alert and busy, but we need not be terrified, and we need to examine the trends with care.

Not all the trends can be directly extrapolated from the experience of the last war. Some of them can be reasoned from inherent possibilities and limitations. The new things to enter are very high-speed bombers flying well above the reach of ground guns, and the very fast guided missile as a weapon against aircraft. Increasing the speed and altitude of bombers does not appear to make the accompanying fighter more valuable. In fact, as all speeds go up, the advantage to the defender, with the aid of his ground radar, seems to increase. For a time, and at any time against a poorly prepared enemy, the modern bomber at very high altitude can proceed with comparative immunity, even though it has a serious problem in hitting anything precisely from its position eight or ten miles up. Against a fully prepared enemy, with plenty of early warning, interception radar on ground and in air, and the best of jet pursuit ships, it would appear still to be vulnerable. Accompanying fighters could help it very little, and its own armament would necessarily be restricted by the very condition of long range and high speed.

Into this situation the high-speed guided missile now enters. It can be used locally, like a gun, or carried to a threatened spot and fired from a plane. It has not the gun's limitation on ceiling and can go as high as the bomber. Its speed of a bullet pretty well guarantees it against being shot down in flight. Jamming it is difficult. It is directed into its target and carries a proximity fuze. To protect an enormous area in this way might be uneconomical. For the defense of restricted areas it promises to be a deciding factor. It can be used air-to-air, but here again the interceptor can use it to better advantage than the bomber. It is a device of the future, but it should come in by the time there are great stocks of atomic bombs. It may well render all mass bombing obsolete when two highly technical, alert, and industrially advanced combatants clinch.

No man, without a crystal ball, can see clearly the future of high-altitude bombing. That is not the purpose of this discussion. But we can dispose of the extremes. On the one hand, those who argue that the advent of the atomic bomb makes no change at all are certainly oblivious to cold facts. When it is possible to place in a single package the explosive equivalent of thirty thousand tons of TNT, a radical change has occurred in the art of war, and we had better not forget it. At the other extreme there are those who argue that on the outbreak of war some time in the future all the principal cities of both belligerents will be promptly and utterly destroyed by mass fleets of bombers carrying atomic bombs, and who, arguing thus, completely ignore the enormous strides being made in defense. In between lies reason. The bomb will be important but not absolute. How important depends greatly upon what happens in methods of defense. They might be inadequate, or they might develop to such an extent as to remove much of the harsh threat the bomb brought to the world. We cannot tell yet, but we should search diligently.

It also needs to be emphasized, before we leave mass bombing by atomic bombs, that there are two periods to consider, one the near and the other the relatively remote future. The race between methods of delivery and methods of defense has just started. As things stand today there is not much doubt that, because of the very high-speed, high-altitude bomber, bombs could be delivered onto their targets with considerable assurance. This is not being argued in detail here. It is the long pull we are primarily interested in. For the long pull the defense certainly has a chance.

We need to consider also the sneak bomber, coming in at low altitude to avoid radar warning and too low to be fired upon effectively by the batteries of guns or missiles adapted to bring down the high bomber, carrying an atomic bomb to place on a target, possibly without much hope of escaping the blast itself as it does so. This is a complex matter, and it would take much space to analyze it at all completely. It is by no means one-sided. The defensive fighter is at a considerable disadvantage in finding the invader. Antiaircraft guns give small coverage, and the time the bomber is in range of any given gun is very brief. Barrage balloons have considerable limitations, except in the protection of important concentrated targets. On the other hand, navigation of a low-flying bomber to its target over unknown hostile territory is no cinch. The planes would presumably come from some distance, and they would have to be relatively large to carry an atomic bomb and have the necessary range. They would not flit about just above the trees, night or day, at slow speeds. They would be ponderous and fast, and high enough for suitable artillery to get at them.

Short-range, rapid-fire guns for protecting ships from hordes of attacking planes were given considerable development during the war. They were especially improved by gyroscopic sights. But that was a vastly different problem: the protection of a concentrated target against many highly maneuverable attacking planes coming in from all angles. The guns were not fully automatic, or radar-controlled, or supplied with proximity fuzes. Costs and complexity were too great, and reliance was therefore placed rather upon volume of fire. For the large sneak plane coming in at low level to attack a city, an entirely different system is needed. It can hardly be described, or even speculated about, without exceeding proper limits regarding secrecy. But if the high-flying plane can be stopped, with ingenuity and adequate matériel and training, so can the one that attempts to sneak in at low altitude. Not every plane, of course, at either altitude, but enough to tip the scales. One does not determine the outcome of a war by placing a few bombs, even atomic bombs, on a few cities.

It is now fairly well accepted that the lone bomber setting out to fly an atomic bomb at high altitude many hundreds of miles over the territory of a fully alert and equipped enemy would never get to its target. The movement of every plane in the country would be continuously charted and followed by radar. Against an unknown plane, pursuit planes would take off in haste and numbers. They could without much doubt bring it down. Perhaps it could sneak in at low altitude, but that would also be risky and difficult. Accompaniment by fighters would not help much. Instead it is now conceived that great fleets of bombers and fighters would fly, and some in the group would carry atomic bombs. This idea also has its limitations.

It takes us back to our earlier consideration. The atomic bomb may be cheap, compared to ordinary high explosive, when measured in terms of the amount of destruction caused, but it is cheap only if its means of delivery is cheap, if one plane can do the work of a whole fleet. If it takes a whole fleet to carry an atomic bomb, then most of the advantage is lost, and we get right back to the question of whether mass bombing pays at all. Of course the whole fleet might carry several such bombs, and proceed to more than one target, but the attrition in penetrating to more than one target would considerably cut this possibility. The advantage is undoubtedly still with the atomic bomb as against high explosive, and the advantage is substantial. But the point here is that the advantage is not overwhelming. The present qualms, harking back to the days of Douhet, do not come from a fear that warfare may be somewhat modified, but from a notion that warfare has suddenly become of a completely different order of magnitude. An examination of the probable nature of future air bombing does not bear out this notion. Great fleets of bombers may not even fly at all, if the defense is strong and alert.

At every point above we have used the word alert, and we need to treat the case of sudden surprise attack—the enlarged Pearl Harbor, the attack against a sleeping nation—and we shall come to it. Also we have to consider subversive methods, slipping bombs into place through an underground organization, or by means of innocent-looking merchant ships.

We must take into account the delivery of bombs by submarine, the bombs being lobbed into coastal cities by rocket projectors perhaps from submarines twenty-five to fifty miles at sea. This is no mean threat; such an attack can be made if an enemy has atomic bombs and submarines that can move into position while submerged, do the necessary aiming, and fire their missiles without detection. If the submarine were generally overcome without question, we could dismiss this aspect, but the submarine is by no means overcome, and the scales have recently tipped in favor of the attack as far as undersea warfare is concerned.

We in the United States have to defend only a relatively few points when it comes to attack from the sea by submarine. We would not defend Miami, for example, much as the people of Miami might object, for the activities of Miami are not essential to the prosecution of a war effort, and the question would be one of national survival when the issue became joined. Mine belts and belts of listening devices would have their place. Radar warning, followed by a prompt pounce by a rocket-equipped plane upon an emerging submarine, would have considerable prospect of success if well organized, for navigation fully submerged is no small task, and a submarine could not fire with assurance the instant it surfaced.

The real solution, however, is to drive the submarine from the seas. This can be done, but not by methods carried over from the last war. The importance of undersea warfare from our standpoint as a nation is fully as great as that of air warfare, and it warrants fully as concentrated and flexible attention. Until we have the problem solved we are not protected against atomic bombs from the sea. They could not alone win an all-out war against us, but they could be a highly significant factor. This is a very technical matter, where the most important phases are necessarily secret and should continue so. The problem can be solved, but it will take years and heavy effort. It will not be solved by some inventor's having a brilliant stroke of genius.

Thus far the atomic bomb has been discussed merely as a powerful concentrated explosive. Before leaving this aspect we need to make one further point. The atomic bomb cannot be subdivided. This is inherent in the physics of the situation. If we have one atomic bomb, with the explosive equivalent of thirty thousand tons of TNT, we cannot split it up into thirty thousand bombs, each with the explosive equivalent of one ton of TNT. Its energy has to be liberated all in one spot, and the total destruction is thereby reduced compared to what it would be if it were more effectively distributed. The destruction occurs in one sudden blast, not in driblets that can be repaired about as fast as the damage occurs, as was nearly the case in some German production areas; but this advantage of suddenness is offset by the necessary localization. There will be no shells from guns carrying atomic explosives, nor will they be carried by marine torpedoes or small rockets or in any other retail way. Atomic bombs will be used only against important targets to

which it pays to devote a large effort, targets that warrant the
expenditure of a large explosive energy in the light of the cost
considerations at the time.

Another question that needs to be considered is that of the
application of atomic energy for propulsion. Will not atomic
energy drive ships and submarines, long-range aircraft, guided
missiles, enormous tanks and whatnot, and thus completely alter
the methods of making war?

It has become evident, in the last few years, that the application of atomic energy to peaceful industrial purposes is far off.
The ultimate importance is still great, but the conception of a
maze of power plants covering the earth, deriving their power
from atomic fission and thus storing and utilizing the very materials that could quickly be converted to bombs, has largely disappeared for a considerable distance into the future. The simple
fact is that costs at the present time are against it; the advantage
of atomic power over power from water, coal, or oil is not great
enough, if it now exists, to warrant extensive use, even though
such use might be justified in certain special cases where power
is highly valuable and conventional power plants are expensive
for one reason or another. This economic consideration would
have slowed down use decidedly, even if the state of international relations had not forced us to keep such supplies of
fissionable materials as we may have readily available for use
in these bombs.

However, it is often asserted that in time of war or preparations for war cost considerations disappear. As has earlier been
said, there is no more troublesome fallacy. Cost considerations
do not then disappear; they take a different form. In fact, in all-
out war it is all the more essential to conserve resources and effort
and apply them where they will produce the greatest effect, so
that costs are then of more real moment than they are in peace.

There are severe technical problems in building atomic-power
plants of small size. In a large plant one can tolerate the heavy
weights of the shields necessary to protect personnel. Obviously,
one would not put an atomic-power plant in a plane if fifty tons

of lead had to be carried to shield the pilot against its radiations. There are also severe problems of materials and temperatures.

But leaving all this aside, there do not appear to be many applications of propulsion by atomic energy that will radically alter the art of war. Atomic energy can undoubtedly be used in time to propel great ships, but the great warship is of waning importance in war, and we do not face a prospective maritime enemy. Atomic energy could probably be used to propel submarines, and this might complicate an already difficult problem. If one wishes to stretch his optimism somewhat in regard to overcoming technical obstacles, atomic power could be used to propel bombing aircraft. But would it be used for such a purpose when the bombers could be propelled by gasoline and the fissionable materials saved for use in more atomic bombs? Certainly not, unless the atomic-propelled bomber could perform an essential function that would otherwise be impossible, and it is difficult to visualize such a set of circumstances.

In the long history of the race atomic power will come to be of great importance, but not soon. Neither will atomic fission soon greatly alter methods of warfare, except by supplying a powerful and compact explosive and by providing radioactive materials. This second matter is the final phase of our subject in this chapter. Toxic materials that can be used to destroy people are produced by atomic fission in two ways—first, during the production processes, and, second, upon the explosion of the bomb. We shall consider the second way first.

The products of any great explosion are more or less toxic, but the explosion of the atomic bomb produces toxic matter of a new type. Radioactive materials have been used for years in very small quantities in the treatment of cancer and for other medical purposes. They act because the rays they give off destroy tissue, and in therapy they have to be used with great care to produce a desired destruction of tissue, but not too much. When an atomic bomb explodes, it produces radioactive effects of enormous magnitude, first, in a great burst of radiation at the time of explosion, and, second, by leaving behind quantities of radioactive materials that persist and continue to radiate. The radiation initially produced by an air-burst atomic bomb is equal to that from thousands of tons of radium; its intensity decreases with time, very rapidly at first and slowly later on. This radiation can kill—suddenly or lingeringly. It can disable without killing.

In the first explosion of an atomic bomb people are killed in three ways: by the force of the explosion, by the intense heat, and by the burst of radiation. Close to the point of explosion all three operate, and the effect is bound to be lethal, except to personnel in strong, well-built shelters. Farther out, the effects diminish. At any substantial distance a very moderate shelter indeed protects against the burst of radiation. For a person in the open the lethal radius of all three is not greatly different, nor, apparently, is it for nonlethal but serious effects. Thus, the fact that the explosion is accompanied by a burst of radiation may increase its killing power somewhat, but not greatly. It makes the bomb a bit more deadly, but not enough so to affect overall considerations strongly.

It is quite a different matter when we come to the persistent radioactive material created by the explosion. This can spread the killing and maiming over greater distances, and it can render areas uninhabitable temporarily or for very long periods indeed. The extent of the effect depends upon how the bomb is exploded. If it explodes high in the air, where it gives maximum blast effect over a considerable area, as it did at Hiroshima and Nagasaki, the radioactive materials are carried into the high atmosphere by the rising column, and the effect on the ground is negligible unless perhaps the materials are brought down in significant quantities by rain soon after the explosion. Even if the bomb is exploded close to the earth, as it was at Alamogordo, the contaminated area on the ground is small and the effect negligible in comparison with other effects of the bomb. But if the bomb is exploded fairly deeply under water, as was one of the bombs at Bikini, it creates a wall of mist that flows out over the water, contaminating all that it reaches, so that some of the Bikini ships received so much of the stuff that decontamination was difficult and laborious. Such a wall of mist could be carried by the wind over a section of a city and make it uninhabitable. There need not be immediate casualties in great numbers if evacuation is prompt and orderly, but delay would be fatal, panic probable, and large areas of the city would be uninhabitable deserts for years. This fact greatly increases the military power of the atomic bomb. It affects the tactics of high-altitude bombers against cities with appropriately placed bodies of water, insofar as greater damage would be done by dropping the bomb in the water than by exploding it over the city, but still leaves some doubt as to the choice even in ideally arranged situations. It decidedly increases the danger of the bomb as delivered from the sea by low-flying planes or rockets from submarines or by innocent-appearing merchant ships. Yet there are limitations: the depth of water must be considerable, more than is often available near large cities, and the wind must be right.

In addition to the radioactive material produced when an atomic bomb explodes there is also the radioactive material that is a by-product of the manufacture of bombs. Atomic piles operate by reason of atomic fission, a chain of atomic explosions, one triggering off the next, producing heat, and accumulating in the pile the explosive material that is then used to make bombs. Just as the atomic fission of a bomb explosion produces radioactive materials, so does the mild continuing fission in the pile. A long process at a slow rate and under control produces heat, radiation, and radioactive substances and also produces fissionable materials by which heat, radiation, and radioactive substances may be released suddenly and at great concentration and intensity in an explosion. The amount of each of the three produced gradually in the making of a bomb is comparable to the amount produced suddenly when the bomb goes off. The heat of the manufacturing process could be used in large power plants. The radiation may be utilized for secondary processes or by-products. But the radioactive material must be disposed of, at great expense, underground, where it cannot get into the atmosphere or into streams; or it can be saved and utilized as a toxic material for war, quite apart from the bombs themselves. Utilizing it thus would be no easy matter. The material cannot easily be gathered, transported, and dropped or projected upon the enemy like a poison gas, for it is deadly at a moderate distance to any who approach, unless they are behind thick shields. But it could be used in war.

There is no logic in considering such toxic material alone. We have long had poison gases, and they were used extensively in one war. We are told today, sometimes stridently, that toxic biological materials—germs, viruses, rusts—are the overwhelming toxic materials of future wars. There are many ways of killing men. They can be hacked to pieces or perforated by arrows or bullets. They can be torn to pieces by explosion. They can be burned by fire. They can be poisoned, and in many ways. There are chemical and biological poisons that kill, and there are now radiological poisons that kill just as completely. During the long and vicious history of warfare this last possibility has always been present—the poison-tipped arrow appeared in the jungle apparently nearly as soon as the arrow itself, and the poisons used were as powerful as any we now have—a surprisingly small amount properly distributed would slay the entire population of the earth. But poisons have never been really important in all-out war; they appeared prominently in one world war but not in the next. We are now told that they have altered the entire concept of war, that they will destroy great populations, that they will reduce the world to barbarism. Indeed they may be of great importance, and we cannot neglect them. Scientific progress, if we call it progress when man distorts the potentialities of scientific advance to his own harassment, has produced new toxic agents having novel applications to war; it has done so in the field of atomic energy, in the production of new and more powerful chemical poisons, and in the field of the biological sciences. But there is much more to toxic warfare than the toxin itself. It must be delivered in effective form to the enemy. We have had powerful toxins for generations; they have been sparsely utilized in war, and the reason has been that the difficulties of application and dissemination have made them competitors, usually on a secondary plane, with other methods of destroying the enemy. We can go further than this. For generations we have had toxins of such power that readily manufactured amounts, properly applied to an enemy population, would be overwhelming, and they have hardly been used. Now we have still more powerful toxins. The question of their use, of their relative effectiveness, still depends upon methods of delivery. It is here that the question rests in regard to future use. The nature of the new toxins themselves widens the nature of the delivery means that are possible, but there remain three main methods, by projectiles, by aircraft, and by subversive methods. If these have altered so that mass delivery in effective manner is feasible, then indeed a future war might become primarily a toxic war. There is no demonstration that this is the case. It is far more likely that toxic methods of warfare will remain merely an alternative, that indeed they will continue to occupy a secondary position. Toxic materials are highly competitive among themselves, but there is no basis for jumping to a conclusion that here now is the absolute weapon that makes all other ways of destroying men obsolete. These materials enter, they enter decidedly, when we estimate the nature of possible future war, but they are more important to our thinking about subversive methods than about open all-out combat between fully prepared adversaries.

CHAPTER X

THE NATURE OF TOTAL WAR

"There is no such thing as total peace, but there is such a thing as total war and total annihilation of rights. Totality, which is inhuman, belongs to fascism, not to us."   —MAX ASCOLI
The Power of Freedom. 1949

WHAT, then, will war be like if it comes again to the world in total form? Will it wipe out civilization and reduce us all to barbarism?

We wish to avoid war in any case, but that desire is a vastly different thing from pondering whether the next war would leave only a straggling, numbed remnant of the race to start the climb all over again for a millennium. We have heard much from the prophets of doom, and their cry is that the situation is so desperate that any alternative to war is preferable, or almost any alternative. But alternatives may well include the abandonment of our liberties, an acceptance of a new slavery as the lesser of evils. It is fairly easy for us at a distance—for the oceans are still there, and we know the day is not yet here when they can readily be crossed in overwhelming power—to reason that stout men would never yield, no matter what the odds. But it is different when the menace is close by: there is a fascination of fear, there is a sinking of the spirit, in standing alone against a terrible and subtle threat without the sustaining exhilaration of the direct clash of arms, and yielding may be the result merely of weariness of the spirit. So we should take a hard realistic look at the subject, for we are deeply and immediately involved, even if the oceans are wide.

To take that look has its difficulties. War plans should not be made in the public press. On subversive warfare there is much that the public should know in detail, for civil organizations and

113

the actions and observations of individual citizens will be the main reliance in coping with it. But conventional war, the operation of the professional armed services, and the plans for utilizing our strength against a formidable enemy if war comes, are something else again. To spread our entire thinking before a potential enemy is senseless; it gives him a very real advantage, and an avoidable one. There is a reasonable balance between keeping our people informed and intelligent and unnecessarily supplying data to a potential enemy.

Since the war ended we have at times erred and departed from this balance. As new aircraft have been developed, we have described their performance in some detail, published photographs of them, and discussed their merits and limitations. When a new guided missile flies well, the whole world knows of its success. We even discuss parts of our war plans, and examine in the full light of public discussion the relative merits of carriers and land-based planes for strategic bombardment, spreading out concepts of how we would use our power for all to read. Doing these things does not make sense, and elsewhere we shall consider why we do them. Here, however, it is sufficient to note that there is a distinct limitation on how far one may legitimately go in examining this question of future war. Still there is a reasonable area of discussion. So much is already well known that the boundaries are wide, and there is sufficient latitude for the examination of the problem that concerns us: whether a future all-out war would be so devastating as to set back radically the clock of civilization. We shall examine it in the light of the technical evolution of weapons, as we have traced it, and in the light of their probable power and interrelation in the future, as far as we can now see. Our examination is incomplete, for it necessarily omits many important factors. It is tentative, for times change and weapons evolve. It is one opinion among many. Yet it is worth pointing out in summary form that there is a probable course of events that runs counter to both the dire predictions and the unwarranted optimism of the two groups of extremists.

We need to examine two situations: one that obtains now,

when we have the only stock of atomic bombs, and the other that will be with us when others have stocks. We need to think of the surprise attack separately, for it is especially alarming in its potentialities. We need also, before we conclude our survey, to consider cold war and subversive war.

We can discuss the immediate future rather briefly. If all-out war came tomorrow, or in a few years, or at any time before our enemy had a considerable stock of atomic bombs, would it destroy civilization? It certainly would not. Is civilization destroyed, after the ordeal of the two world wars it has experienced since 1914? It has undoubtedly been sorely harassed. The misery has been so great that we shrink from thinking of it fully. It has been folly for the race as a whole on this earth to have spread disaster and to have wasted its substance. Undoubtedly we should have been further along in general standards of living and in creature comforts; probably we should have advanced further in the arts; possibly but not probably, science, including the conquest of disease, would have been on a sounder basis for the future if there had been no lust for conquest and world domination. Certainly it would have been a happier world if masses had not been taught to hate, if international morality, for what it used to be worth, had not been abused and perverted. Yet for all this, civilization was not destroyed in the last generation. It was made to divert its path in unclean ways, it was distorted, but its march was by no means stopped.

Neither would another early war stop it. Such a war would be a tough slugging match. Intricate techniques would enter, and some of them would be new, following from the trends of the last war. But it would be no affair of pushbuttons.

The opening phases would be in the air, soon followed by sea and land action. Great fleets of bombers would be in action at once, but this would be the opening phase only; that is, unless our air fleets carrying atomic bombs forced a decision at once, and this is not probable. They could undoubtedly devastate the cities and the war potential of the enemy and its satellites, but it is highly doubtful if they could at once stop the march of

great land armies. With a severe loss of supplies, with their
transportation harassed, the enemy armies would be less than
all-powerful. They might indeed disintegrate seriously, and undoubtedly they could be overcome. But to overcome them would
require a great national effort and the marshaling of all our
strength, for they would be far off and Europe has not yet recov-
ered. The effort to keep the seas open would be particularly
hazardous, because of modern submarines, and severe efforts
would be needed to stop them at the source. Such a war would
be a contest of the old form, with variations and new techniques
of one sort or another. But, except for greater use of the atomic
bomb, it would not differ much from the last struggle.

While it went on would our cities be devastated? Would great
fleets of bombers or ocean-spanning missiles guided unerringly
to a target flatten our homes and factories and kill millions? We
should undoubtedly suffer some of this sort of thing. Our coastal
cities in particular offer tempting targets for bombs lobbed in
by submarines off the coast. We should be the primary target, as
far as we could be reached, for we should be the most powerful
adversary. But in all probability we should take no such battering as England took in the last war, and probably not enough
to weaken greatly our war potential.

In the first place, as we saw earlier, there is no such thing as
an ocean-spanning rocket or guided missile capable of precisely
hitting a target in another continent. There will be no such thing
for a long time to come, and even if there were it is exceedingly
doubtful whether it would be worth its cost if it carries conventional high explosives. By the time there are intercontinental
missiles, if ever, we shall have an entirely new set of circumstances to consider. In discussing an early war we may forget
such weapons, even if those charged with their long-time development can by no means neglect them completely, for science
progresses and techniques alter.

Would great fleets of bombers carrying TNT, or its improved
relative, RDX, blast us down? There are no such great fleets
today capable of spanning the requisite distances and returning,

carrying heavy armament to protect themselves and bearing loads of bombs. From relatively near bases such bombing could of course be done, and on some critical targets it would pay well, but we should be weak indeed if we allowed such bases to be established near us and used. Refueling in the air, which we have accomplished and also publicized, will extend range for a few bombers, but is enormously difficult to carry out for large fleets. We need to consider principally great bomber fleets that can span many thousands of miles without refueling, ready to fight their way to a target, carrying heavy loads, and to return. In spite of the remarks of those who delight in ignoring some of the teachings of physics and chemistry when they paint the glamorous picture of aeronautics of the future, there will not be such fleets for a long time to come. Will they, however, fly only one way, drop their bombs, and then abandon their craft? For a demonstration raid, such as we put on over Tokyo, per-haps they may, but this cannot be really devastating. If they are carrying ordinary high explosive, the cost of one-way raids is altogether too high.

There is such a thing as conservation of force in the conduct of war. It still holds in spite of overenthusiasm, and a combatant disregards it to his peril. For a war that is fought to a finish, any method used to bring down the enemy by destroying his will to fight or his means of doing so must do him more damage than it costs. It must be part of a sustained effort, or it will often do the greater harm to the fighting power of him who launches the weapon. There are exceptions, of course, especially where the psychology of war enters. Thus the German V-2's may have been worth the effort on this basis; they were certainly a failure in the physical damage they caused when balanced against their cost, especially when the same effort might have produced enough jet pursuit planes greatly to impede our bombing and our land offensive.

This cuts both ways, of course. We need to have in absolute readiness the air fleets to deliver atomic bombs if war comes. We need to have planes to hunt submarines and to seize bases and to support land offensives. But do we need, in addition, great fleets of planes to carry ordinary high explosive for the mass bombardment of enemy cities? The effectiveness of this sort of effort was a borderline matter in the last war. Now our costs of planes, per pound, have been multiplied by ten, as speeds and intricacy have mounted, and the distances we now consider are extreme. It is highly doubtful if a case could be made out for them.

We have one thing more to consider. Would fleets of planes be carrying diabolical warheads of one sort or another, and would they warrant the effort by their appalling damage? There is need to consider poison gases, biological materials, and radioactive products. The last can be disregarded until a potential enemy is manufacturing bombs, but they could appear as a by-product of manufacture before there was a stock of bombs in actual readiness. These three weapons are alternatives, and they are subject to many of the same limitations, centering largely on the problem of dispersal. We can hardly study all the intricacies of this subject. Perhaps the relative standing of the three is indicated to some extent by the fact that all three are being studied, as is well known, and the effort on gases, measured in dollars, is as great as or greater than what we are spending on the whole question of biological warfare. Certainly war gases, even if they were not used, stepped ahead in general deadliness during the last war, especially because of German efforts. Nevertheless, more than a mere strategic question was involved in the neglect of them in the field, more than the question of which side had the advantage. There was also a question of tactics. The first war lent itself rather completely to gas warfare, with stalemate in the trenches and bodies of troops in long-continued deadlock. Distribution even then was principally by shells, mortars, and the like, even though low-flying slow planes were often in a position to spread gases in much the same way that an agricultural plane dusts a field. It is a vastly different problem today to distribute war gases or any substitute over great distances.

Bombing by slow planes as we knew them in the last war is

now probably obsolescent as between fully prepared and highly technical enemies. Radar, the proximity fuze, and above all fast pursuit planes took care of that. At short distances, on the deck, individual planes might well penetrate, as was done on occasion over Germany by the British Mosquito fighter-bomber—the versatile plane that also was often used for night forays at twenty-five thousand feet. But great fleets of low-flying slow planes would be easy marks for an enemy ready to receive them. Very high-speed bombers at great altitudes are different; they can be above antiaircraft fire, and relatively immune to the pursuit ship, except under circumstances of excellent warning and control. But the high-flying plane is going to have a hard time hitting anything whatever with its bombs. It certainly is not going to fly about leisurely sprinkling an area with gas or anything else. Of course it can deliver its bombs, and these may have other things than high explosives in them; but the limitations on this method are great, for gas is an area weapon.

The war gases have improved, but the means of delivering them have become more difficult. It is hard to say whether in the last analysis they would be used instead of high explosive or incendiaries. It is harder still to tell whether biological materials or radioactive substances would be preferable. There is no sign whatever that any of them has become the ultimate weapon, the device so deadly that it makes all other methods obsolete. Where there is a close difficult choice between methods, neither is overwhelming in comparison with the other. If mass bombing using high explosives, considered apart from the movement of land armies, is now of doubtful utility, or even obsolete, a close choice of a different type of bomb will not swing the balance far. Though the long future may of course hold in store other things, we are dealing now with an immediate or moderately distant future. We can by no means neglect these instrumentalities, but there is no occasion to become overexcited about them.

While we are about it, the high-flying bomber warrants additional comment, for much of popular thinking revolves about it. Nor is this a one-sided affair; very few technical questions of

warfare are. It is popular to speak of fleets of bombers flying in
the stratosphere and at supersonic speeds halfway around the
earth, without intermediate bases or refueling. True, we can
undoubtedly build great bombers that fly very high and very
fast, above the speed of sound if we please, although we are not
doing so yet. We can break the sonic barrier. But we cannot
break the general law of physics that the faster one goes through
any medium the greater is the resistance. In other words, if we
want high speed we have to sacrifice range or make the craft
very large, and incidentally very expensive. But, say the confirmed optimists, we shall soon have much better engines that
will take care of all that. Perhaps we shall. But the development
of engines has had a lot of attention and has been a gradual
evolution over many decades. Jet engines are not more economical than reciprocating ones; they are less so. They may sometime catch up, or even lead in fuel economy, but the chance of
their making a radical jump that changes the whole nature of
the problem seems to be remote. Let us leave out of consideration here the matter of driving planes by atomic energy; it is a
highly controversial subject and a highly technical one, and
there seems to be agreement only on one point: that it is at least
a very long time off, from any practical standpoint. The chances
are decidedly that, as we push up bomber speeds and altitudes,
as we try at least to maintain range at the same time, our planes
will grow larger and larger and more and more expensive. There
is such a thing as the law of diminishing returns, even in warfare. The resources of this country, or of any other, have their
limits. If planes are forced to high altitudes and speeds to secure immunity there will be fewer of them.

Will these high-flying planes be so completely immune? For
a time, yes; but, looking ahead, perhaps not. For we have to
think of the time required before atom bombs in quantity are in
the hands of two belligerents. The future enemy of the high-
flying bomber is the guided missile. The guided missile had its
baptism in the last war; a few guided bombs sank ships and
cruisers in the Mediterranean, and we used them in eliminating

the remnants of the Japanese Navy. These uses were merely a start. Though the idea of practical intercontinental missiles of this sort may be fantasy, short-range devices are not.

The general idea is simple: wherever it is not possible or desirable to use a man, substitute a robot. Either give it orders from a distance or endow it with sense to respond automatically and seek a target. There is no dreaming in this; it has been done in a dozen ways. An instrument can go where a man cannot. It can stand high acceleration or heat or cold. Its judgment is infallible on the simple matters it is built to encompass, if it is in good order. The essentials for such a robot, to battle the high-flying bomber, are already available.

One of them, the most powerful device of this general nature in the last war, is the proximity fuze. As then used, it did not give guidance, but it blew itself up at the moment when it could create the greatest damage. It multiplied the effectiveness of large antiaircraft batteries by five or ten, and introduced about the same factor in artillery when it was substituted for timed fire. It was an extraordinary device, and its great success marked the way for a host of applications. The extraordinary thing about it was that a radio set, compressed into the size of a small baking-powder can, could be fired off in a gun, experience forces pressing upon it with a ton's weight, and still operate so subtly that it could detect a plane in its vicinity by radio reflection. It showed that all bets are off when it comes to using mechanism instead of fallible and delicate men to control levers and switches on fast-flying vehicles.

Another vital possible component for the robot weapon we have already mentioned is the ram-jet, the simplest of all engines. It is a hot flared tube, no more. It is of no use whatever at low speeds—its fuel consumption is high in any case—but to drive a vehicle for a short time at very high speeds indeed, it is unmatched. The ceiling of artillery, with any improvement that seems reasonably in sight, may prove to be well below the ceiling at which pressurized planes can fly. There is no such limit to the ceiling of a missile that carries its own ram-jet to push it along. If the plane can find air for its engines, so can the ramjet. It can attain speeds of the same general order as those of bullets.

Now, as the proximity fuze suggests, this missile can carry a device that will sense a target in its path, steer the whole assembly in its direction, and crash into it or explode when within lethal distance. And this missile does not need to be highly expensive, certainly not in comparison with the cost of the modern aircraft it aims to bring down. It can be launched from the ground or from a plane. It is not here as yet. The need for it is great, however, there are no physical laws that set prohibitive limits on its performance, the elements for constructing it all exist, and it will appear.

Any kind of bomber whatever, full of men and bound on a mission, might well have a thin chance if it ventured into a nest of these hornets. It could neither see them nor dodge them; they come too fast. It might jam their controls, if it knew they were coming, and if it also knew what kind of control to jam, but the odds seem to be heavily on the missile in such a contest. It is a devilish device. It may yet bring a feeling of relative security to the world.

If war came right away, such devices would play a part only if the war were long. For war in the middle future, they may be decidedly important. If war begins then, and the opening settles into a slugging match, the ultimate advantage will lie with that nation which has scientific and technical ability widespread among its people, industrial capacity and versatility, and a determined will to prevail. It is here that our strength lies, and it is in these aspects that we must preserve it.

There is little danger of all-out war in the near future. If it comes it will be by miscalculation, not by design. All peoples are weary, and our large wars come only after intervals, not so much because men forget as because they recover. But there is a greater deterrent than this. Even apart from the atomic bomb the strength is ours, such strength as the world has never seen, such strength that the ultimate outcome of an early war could hardly be in doubt. War would again be a brutal contest, evil and distressing; the damage would be large, and we should share in it, but we would by no means be knocked out, and we would win it. The whole world knows this. Unless we get soft, unless we are clumsy in meeting the tactics of cold war, we are not in immediate danger.

But what of total war sometime in the future between two great nations, each one of which has a large stock of atomic bombs? Before we plunge into a discussion of this matter it will be well to examine closely what we mean. The problem is sometimes oversimplified by assuming that in a few years both we and Russia will have such stocks that we will then inevitably fly at one another's throats, and that within a short time after war opens great fleets of bombers will completely devastate all the great cities of the earth, thus ending the war with both contestants reduced to impotency. It is not nearly so simple as that.

In the first place, by a large stock of atomic bombs we mean such a stock that, after the inevitable losses in delivery, inaccuracies in aiming, and faulty functioning, enough could be delivered on targets to destroy substantially the warmaking potential of the enemy. How many is that? It is absurd merely to count enemy cities, multiply by the number of bombs to destroy each one, and add to a total. It is also absurd to take the million-plus tons of high explosive that fell on Germany (which did not stop Germany from fighting vigorously on land against armies that outnumbered it in active divisions) and convert this to an equivalent number of atomic bombs in terms of destructive energy. To do these things is to ignore the defense, and the defense is mounting and will mount far before we arrive at such a condition as we now contemplate.

. To build a large stock of atomic bombs is an undertaking that will strain the resources of any highly industrialized nation. The strain will be greater if very dilute sources of raw materials have to be utilized, as seems probable. To build the fleets of bombers that can deliver them over thousands of miles, with all the attendant paraphernalia of a great air force, will also strain resources. But to do only these things, as the time of trial by arms approached, would be foolhardy. There must also be a defense system. It will consist of radar nets for early warning, other radar nets for ground-control interception, communication systems, landing fields, pursuit aircraft, guided-missile stations, spread over enormous areas and concentrated around hundreds of cities. To build this system would also strain the resources of a great industrial nation. And when both potential contestants have accumulated all three—the atomic bombs, the bombers, and the defense system—what do we find? Perhaps that the attrition caused by defense measures is so great that the originally estimated stockpile of bombs is by no means enough. So a new program would have to be undertaken with higher objectives, and the race would go on in an ascending spiral.

This is not to state that it is highly improbable that the conditions stated in our premise will ever occur, for, unfortunately, we can take comfort in no such hallucination, but rather to indicate that the time when we shall have to face the issue of two groups frowning at each other over adequate atomic-bomb stocks, and all that goes with them, may be very far off as time is reckoned in present international relations. It is a far cry indeed from the time when the enemy has a bomb. Before the time comes when a war would be primarily an atomic war, many things may happen. We may be living by then in a different sort of world.

In the interim it may happen that one potential belligerent will have an adequate stock (in the sense used above), as well as all the planes and the defense system that go with it, before the other does, and this may produce a one-sided atomic war, with no other instrumentalities necessary for a conclusion. If such a situation is to arrive, we should be very sure that we get into condition for atomic war first. It seems highly probable that we shall. We do not need to estimate, by extracting all the possible data from published material and ingeniously combining them, how many atomic bombs we now have, and jump to conclusions as to whether they alone are now adequate to bring

an enemy to his knees. To ignore enemy defenses in such an estimate would be naïve indeed. But it is certainly not stretching
a point to assume that, as things go, we shall arrive first at the
condition of an adequate stock for the purpose. If we do, will
it then mean war? If a bellicose state foolishly forced us into
war, by attacking our troops or our vital interests, it would,
but it is improbable that the bellicose state would commit suicide in just that manner.

The real point is whether we would then force a war ourselves,
a prophylactic war, striking while the advantage lay with us and
before it reversed, or before it turned into a condition where
both contestants would be horribly damaged in the first clash.
A dictator would do so. A government truly responsive to the
will of a genuinely free people would not. Certain it is, if we
did thus embark on a prophylactic war, we should lose the very
freedoms we cherish and for which we would be fighting. For
freedom would have to be suppressed in some manner to cause
us to embark on armed conquest of any comprehensive nature,
and this would be armed conquest of the worst sort.

There is another reason why we would not strike, even when
we had arrived at a point where we felt we could do so and
win readily, and with only secondary damage to our own cities
and population. It is our national characteristic that, if we arrived at a point where we fully believed we were in position to
strike and prevail quickly, we should also be fully convinced
that we could easily retain the advantage and widen it, and so
we would wait. We would not strike, because our moral sense
as a free people would not allow us to do so; we would not strike,
because we would sense that we would lose our birthright if
we did; but we also would not strike, because we would be just
cocky enough to believe that we did not need to.

No all-out war will be based on the possession of neat stocks
of atomic bombs adequate for a decision, unless we lose the race
toward that position of adequacy. For if we win the race and
keep it won, war will not follow from that cause. Wars are
brought about in many ways, but what we are here examining

is whether they are likely to be brought on by the existence of the bomb itself.

There is, however, more than one way of losing the race. We have not gone far in it yet, and we already feel the pinch. The race can be lost, as all long races that depend upon man's endurance can be lost, either by doing too little or by trying to do too much too soon. It will profit us little to have stocks of bombs and planes and then to bring our governmental and industrial systems crashing down about our ears. This is a long, hard race we are embarked upon; we had better settle into harness for the long pull and mark well how we use our resources.

Let us be specific. We are building bombs, we are building planes to deliver them, but we are not putting all-out effort into building a defensive radar net at the present time. We are experimenting with it in selected areas, developing elements for it, utilizing war matériel for training and protection of bases, but we are not committed to full building of a great defensive network for early warning and control, spread over the whole country, for the purpose of protecting our great cities. We would be foolish if we did so at this time. The estimates on these matters need to be very cold-blooded. Our situation differs markedly from that of Europe. At present it is evident that the work of a man for a day, the use of a barrel of oil or a pound of copper, can do more to place us ahead in the race if it is devoted to bombs and the means of delivering them than if it is applied to home-defense measures. This will be true for a long while, even after the time arrives when, if the storm broke, we should get some atomic bombs on our cities. If at that time we tried to make our cities utterly immune we should certainly lose the race, for to seek utter immunity would take all our resources for that purpose alone, and even then complete immunity would probably not be attainable. It will take resolution and calm thinking to hew to the line if that time comes. It will take a highly effective system of national military planning, a far better one than we have now.

In all this discussion of defense we have to remember that it

is a matter of degree and of area. The defense of a highly industrialized, concentrated area is a far cry from the defense of a
huge, sprawling nation. There is little doubt, on the one hand,
that by sufficient expenditure of material and effort a strong
point can be protected against aerial attack in the future we
consider, so thoroughly protected that to penetrate the defenses
would demand an effort greater than the value of penetration.
There will be difference of opinion as to how tight such a defense could be; yet the technical trends we have reviewed indicate that it can be more complete than similar situations in the
recent past. On the other hand, there is little doubt that any
attempt to apply such an elaborate defense system to an enormous area would be exhausting in its expense and could hardly
be so successful as to confer complete immunity even then.
Between these two extremes are all sorts of intermediate cases.
The important points are, first, that we should never become so
obsessed with a defense system as to invite disaster by relying
upon it to the detriment of retaliatory striking power, and,
second, that such defensive measures as we take should be employed against specific, real threats and undertaken with proper
timing.

The major effort should go into bombs and their delivery, until
logic shows that a substantial defense effort should be added.
Logic at some time will give the warning. The point will be
reached at which investments in defense installations will so
reduce the ability of the enemy to reach the target that this
indirect effort will pay higher dividends than a direct one. It
will make it more certain that if war came we should win
quickly, and at a minimum net damage to ourselves from every
cause. To estimate the arrival time of that point early enough
to launch programs needing several years for consummation will
call for pre-eminent organization for intelligence and analysis.
We do not have such organization now, and if we value our
safety we had better get it. It will not do to bungle through, in
the type of contest we are now engaged in or in the kind the
future may produce. In the meantime, of course, we should press

research and development on every phase of defense, on radar, guided missiles, and every other element of importance, and especially on their integration into the systems in which they are to work. Research and development cost money, but their cost fades into insignificance compared to that of the construction and maintenance of great defense systems, alert and in effective operating condition, under highly trained personnel. We should not scatter our shots by building a great defensive network now.

What would be the nature of all-out war if it came at a time when great belligerents faced each other over adequate stockpiles of atomic bombs, capable of reducing both to relative impotency soon after the storm broke? The condition may never arise. Before it can come about, there is another type of contest: the race to be prepared, in which we are now engaged. If we lose that race decisively we alone shall be devastated and there will be no atomic war between substantially equal contestants. We have to win that race. Worse than that, we have to stay well ahead at all times as the race goes on. If we do, there need be no atomic war of the fully devastating sort we study. The race will be decided by which system of government and industry is the more efficient and has the better staying power.

, In many ways it is unfortunate to be forced into such an armament race, but is it really disastrous? Mind that the most effective progress in the race will not come from throwing all of the country's resources directly into the making of weapons of war. In fact, that would be a sure way of losing in the long run. To win the race we must have a healthy people. We must raise our standard of living so that more of our population may perform well. We must learn to make our industrial machine operate smoothly and avoid the interruptions because of quarrels over the division of the product. We must learn to avoid inflation and depression. We must somehow produce governmental machinery that will operate efficiently for its intended purposes, so that the selfish interests of groups or sections cannot drain away our energies. We must establish justice and good will among our people and among the races that make up our population, so that our progress will not be halted by internal friction. We may not accomplish all these things, but we should accomplish most of them if all our citizens realize with full clarity that the alternative is someday to enter an atomic war on the losing side.

The challenge is enormous. When war became total the armament race took on a new form. It is now a race not merely for the quick possession of a few battleships or other weapons that might decide an issue, but to attain immense strength of every sort, such strength that the appalling costs of preparation can be paid without wrecking the system that produces them, such mounting strength that armed prosperity can proceed to more arms and more prosperity. In such a race this country need not suffer. It would be well, of course, if all the energy, all the patriotism under the pressure of enlightened public opinion, could be exerted toward well-being alone, leaving the arms out. We know full well that it could not, our nature being what it is. Without the threat and the stark necessity, we should not build our strength to an optimum extent; we should in one way or another dissipate much of it in internal controversy and the juggling of men and groups for power. Even under the threat we may not do too well—we shall certainly be far from perfect in our single-mindedness, our efficiency, and our patriotism— but, fortunately, it is not necessary for us to be perfect; it is merely necessary that we be better, much better, than our adversary. We ought to be able to do that. If we do, the threat of possible atomic war loses much of its terror.

Before we leave this chapter on the nature of possible future total war, there is one more matter to consider. In all that has preceded we have written many times of "an alert enemy," or "a prepared and alert belligerent." What of the beginning of a war when we were not alert, when we were caught off guard, when we were wide open to another Pearl Harbor, this time on a really enormous scale? There are two aspects to this subject; we shall come later to the one involved in subversive methods of making war. Here we need to treat the surprise attack as a part of war in the open, for one thing is certain: if another great war is started by a dictator, it will be opened by a smashing, great, surprise offensive calculated to paralyze us before we are aroused.

There is no question that we face here a handicap, and that the development of modern techniques has made it far more serious than it was before war became altered by the deliberate application of science to its methods. We could not open a war in this manner, but our potential enemies could. We could not, for the ponderous machinery of democracy does not work in that manner. Congress alone can declare war, though the President has the power to repel attacks. Even so, we still could not undertake a successful surprise attack, even if convinced that war was inevitable and imminent. We live in a goldfish bowl in this country. All our moods and passions are open for the world to see. A great democracy cannot be ordered into war; it cannot be taken into war by its elected representatives until the people at large are convinced that we must fight, that there is no other way out. No President would wish the absolute power to order a surprise blow, he would not use the power if he had it, and if he did the blow would not be a surprise. Even a sudden blow by air fleets is not started in a moment, and the preparations in a democracy would be obvious.

In a dictatorship it is different. A surprise blow on a small scale would not be staged even by a dictatorship. It would be large, or it would not occur, for arousing a people fully by a small treacherous assault backfires. Those who study war have realized this fact since Pearl Harbor, if they did not know it before. But a dictator, if his control were tight enough and his iron curtain utterly impenetrable, might stage a large surprise opening when he decided to go to war.

He would send his submarine fleets to sea and place them on strategic trade routes, and this preparation would give him a heavy advantage in the undersea warfare to follow. He would mobilize in secret and set his armies in motion so that they would assault strong points before these could be fully manned, and would smash ahead by reason of their momentum. Above all, he would launch his air fleets, and, if he had atomic bombs, he would direct them at key cities to obtain a paralyzing effect and a full use before attrition by the defense cut down their effectiveness. It is a handicap we must face, and it is a severe one.

It alters the margin we must keep in the race of preparedness. If we assume, as I think we may, that the world is not now as likely to be thrown into open war by accident as it used to be, and that open war will come only if our potential enemy feels it is inevitable and will hence launch it at the most favorable moment for himself, then the attack will come when he feels that his inherent strength plus the surprise factor is sufficiently great to overcome us, and if he feels that time thereafter will be moving against him. He will not launch it unless he feels he can prevail, but the margin against him in normal strength may be considerable and yet he will still feel he can conquer if the advantage of surprise is great enough to overcome this margin. Hence in the race, not a race in armaments alone, but a race in national strength from every aspect that contributes to the waging of war, we must seek a substantial balance in favor of us and our potential allies.

We can do more than this. We can decrease the value of surprise by staying alert, to some extent even in peacetime. We cannot reduce the surprise factor to zero, but we can cut it heavily. This does not mean that we sit with every post manned and every gun cocked throughout the years; to do so would be overly expensive and our efforts would be better placed elsewhere, for we should slacken in time if we tried to be thus continuously vigilant. Short of this extreme of having every man poised and at his post, we can do much.

One element of our strength should indeed be ready and straining at the leash. That is the retaliation force. It is the force that would strike back within twenty-four hours of the time the first bomb fell, remorselessly, through every obstacle, pressing its attacks home, before enemy defenses were working smoothly, for they will be less effective then than later, even if alerted.

The planes of the retaliation force must always be fully equipped and ready to fly, its bombs must be ready to go off, its every element must be so protected that the most severe surprise bombing cannot impede its launching. Its crews must be highly trained, fully briefed, and tested by frequent exercises. There should be an inspection system independent of the line of command up to the very highest echelons, to ensure that there is never any slackening in this picked group or any false assurance. The members of the force should be young, and rotated in duty, for men cannot stand the strain of being thus poised for long periods. But, in comparison with our full military power, the force may be relatively small.

There must be no confusion in the line of command by which it would be put in motion, so that it could by any possibility or mischance either be put in motion prematurely or fail to go into full, unhesitating action on a few hours' notice. There should be no delay in the notice itself. Things will be tense, and organizational machinery should be ready to operate to give the notice in a day if possible, in a few at most. Every day will cost enormously. Here is something we can really do to protect ourselves, and it does not cost much in times of peace.

The President under the Constitution has the power on his own and without Congressional authorization to employ the armed forces of the nation to repel sudden invasion or attack, as he did at Pearl Harbor. Moreover, the President under the Constitution is the judge of what constitutes an attack. Treaties, entered into by the President and approved by the Senate, define to some extent what sort of attack we will consider an attack upon us for this purpose. In particular they can, and do, explicitly state that an attack upon one of a group of countries with which we are in alliance will be considered an attack upon all, and hence upon us. Beyond this is an undefined area where the judgment of the President must prevail, in considering an attack such that his instant action is needed before consulting Congress. If there is time, any President will prefer to consult

Congress before acting. Congress has the power to declare war; in any case, when the President has acted because of attack as defined in treaties, or evidently within the scope of his duties and calling for instant action, it is morally bound to do so. If he exceeded his proper powers, stretched definitions, or acted hastily and unwisely, Congress might indeed call a halt. In the case we are considering we do not need to be concerned with borderline questions. If there were an atomic attack on our lands or overseas garrisons, or on friendly nations with which we were in alliance for mutual defense, the case would be utterly clear and the need for prompt action fully evident. The President, as Commander in Chief, would act at once, and the Congress would without question back him up, declare war, and place the full resources of the country at his command for its vigorous prosecution. From the President down, then, we must be sure the lines are utterly clear and in order, and it will be wise if we attend to this fully in times of peace.

For one thing, the machinery must operate even if the President is disabled, and he might be if there were a sudden atomic attack on Washington. There should be two channels, both of which transmit orders in parallel, to guard against failure or misuse of one, with checks on openness and reliability repeated at intervals, like a trial audit. They should focus on one general, and when he has his orders, in unmistakable form so that he can be sure he has them, he should also have under his direct command every appropriate element necessary for his task, and assurance that when he needs aid those who can render it will have positive direct orders to do so, with no possibility of interference, as he proceeds against predesignated targets. There is, incidentally, no danger whatever that he will move before he has his orders, if he is sure the lines are open so that he can receive them. Generals who come up through proper training do not make such mistakes.

Now we do not need to set all this up tomorrow. No other nation will have an atomic bomb tomorrow, and only after an enemy is ready thus to strike shall we need fear surprise attack of the sort that this system is designed to cope with. But we had best be about it, for the time is coming when we shall need it. We have part of the striking force now, of course, and primary attention should be devoted to being sure it has the tools and is well able to use them and that lines of command are prepared and clear. These lines have been recently improved, but they are still far from perfect.

We wisely placed the development of atomic energy in this country in the hands of a civilian commission, and we shall undoubtedly keep it there. This step was advocated by the Secretary of War at the time and is still supported by the National Military Establishment, in spite of occasional allegations to the contrary. The commission is now charged with the duty of building atomic bombs, stockpiling them, and turning them over to the armed services on orders from the President. When it is fully evident that the services are properly organized to receive them, not before, of course, the store of bombs might well be made the responsibility of the armed forces. There appears to be no possibility of delay because one group now has the bombs and another is to use them, but transfer in advance appears logical when we are ready for it or when the necessity for use appears imminent. If the bombs are turned over to the services, there should be a watchdog somewhere to be sure they are kept in prime condition, and the commission can perform a needed service in this capacity. Yet the exact line of division of responsibility is not the important thing; the real essential is to be sure that every step is clearly the responsibility of a capable individual, and that he knows it.

This is not the only thing we can do to decrease the potentiality of the surprise attack. Later on we can build defense systems and contrive to have them alert and ready. In the meantime our entire military system can be ready, even if it is not at all times acutely alerted, and as affairs become tense there can be means for placing it in this latter condition before a blow falls. We can build a civilian defense system and see that it is ready to cope with disaster.

The principal element of our preparation for possible surprise attack is an intelligence system of high effectiveness, capable of warning us clearly if an attack is being prepared. No iron curtain is utterly impenetrable. The operations necessary to set in motion a major surprise attack are ponderous and far-reaching. Dictatorship and oppression produce individuals who dare to flee and then dare to talk. There is no reason why we should not know, reasonably well, what is afoot; and if we do, much of the value of surprise is lost. But we are not going to know enough as things now stand. Since the war a Central Intelligence Agency has been created, but it has been under the command of military men whose careers lay elsewhere, it creaks at the joints, and it has not yet amounted to much. For one thing, scientific intelligence is not conducted well by Mata Hari methods or through agents who know no science, and there is just as much danger of placing scientific intelligence in the hands of those who do not understand as there is in placing any other part of science in the same tender care. We need a modern intelligence agency in every sense of the word, using modern methods as they were partially developed during the last war, not a musical-comedy affair or a stodgy refuge, not even the half-successful affair we now have, but an organization qualified to meet our needs in this kind of world. It can cut down the threat of surprise attack. It does not cost much; by all means let us have it. We ought to know how to build it, after the experience of the last war, for we did not do badly at all on intelligence work then, after we got our hand in and learned how to do it. But the really able men who functioned then have largely scattered into civilian life, the type of ability needed is rare, and the work is not attractive. The task can be done, by an individual of great mental and organizational capacity, having ample authority and the full backing of the President of the United States. As we value our peace of mind we had better be about it.

What will be the nature of future total war if it comes again to the world? It would be highly technical, and it would be fast and furious. It would leave the world shaken and broken. It would cost millions of lives and exhaust the accumulation of treasure of many years. It would not destroy civilization, any more than the last two wars have destroyed civilization, but it would assuredly set it back.

It need not come if we fully maintain our strength. It need not come if we realistically enough and with enough determination resolve that it shall not. It need not come if we really learn to make our democracy work. It need not come at all, for if the strength of free peoples prevents it for a generation, that same strength can then produce a new sort of world in which great wars will no longer occur. For this consummation we face a task that will test us as we have never been tested before, that will test whether we really mean it when we say that we believe in human dignity and human freedom, whether we can really submerge selfishness and petty motive, and bring our enormous latent power to bear, to make our way of life function with true effectiveness for the good of all.

CHAPTER XI

SUBVERSIVE WAR

"As Americans you have not only a right but a sacred duty to confine your advocacy of changes in law to the methods prescribed by the Constitution of the United States—and you have no American right, by act or deed of any kind, to subvert the Government and the Constitution of this Nation."
—PRESIDENT FRANKLIN D. ROOSEVELT
to the American Youth Congress

FOR A ROUNDED VIEW of where we stand and where we may be going, we need to consider subversive war, for that too has its terrors and causes its unreasoning fears. It is indeed something to be afraid of, as is all war, or for that matter any catastrophe, and we have to order our lives in the light of the terrors of nature and of man, whatever their form may be. But ignorance or half-truths are dangerous; it is fear of the unknown or partly known that usually causes panic, and especially the fear of being trapped by the inevitable. Men who fear do unreasonable things. So it will be well to examine carefully, not to gloss over the real dangers, but perhaps to see them more clearly for what they are.

The threat of subversive war appears in several forms, and we need to consider them separately. First is the form that really constitutes an extension of cold war—political, economic, and psychological activities of a subversive nature to create internal chaos in a possible adversary, to disrupt him or take him over, giving way when these are exhausted to more direct forms, such as poisoning his leaders or his citizens, or decimating his herds or crops by smuggled disease, or introducing debilitating disease into his water supply, as a further means of bending his will. Second is the form in which subversive activity is an adjunct to all-out war—the cloak-and-dagger activities of agents striking an enemy behind his line. Third is the surprise attack, as an opening phase to all-out war, with all these efforts sharply increased, together with atomic bombs smuggled in by innocent-appearing ships, to be detonated at the chosen moment. Before we examine these forms in turn, it is desirable to look at the tools, and biological warfare is the most fearful of these for this underground purpose. We need, too, to consider motivations.

There is no doubt that modern instrumentalities open up enlarged opportunities for all skulduggery of this sort. The progress of chemistry and biology has produced poisons of enormous toxicity, and, worse, has provided means for making them in large quantities. The poison gases of World War I were far exceeded in deadliness by the gases developed during World War II. They are also now joined by the radioactive toxins, deadly agents that make an area uninhabitable briefly or for very long periods, such as contaminated the ships at Bikini, and these can be manufactured as a by-product of the atomic bomb. Then, too, the progress of biological science, with its unfolding knowledge of the mystery of life, teaches us not only how to avoid or cure disease, but also how to create it. So there are hormones that overstimulate and thus destroy plant life, rare viruses that can go on a rampage among cattle, and even the possibility of new diseases of man of which he now has no knowledge and for which he has neither inoculation nor immunity.

The list is formidable, and it will undoubtedly be extended further. The biological sciences in particular seem to be on the verge of marked advances comparable to those in the physical sciences during the last couple of decades. A dam is about to break, and the accumulations of piecemeal knowledge are to join in a great flood of new understanding. The mystery of those clever regulators, the enzymes, which steer and further the chemical processes of organisms, is becoming clearer. The ways in which proteins, the great building blocks of nature, mold the raw materials about them into their own image, producing replicas and hence more proteins from the amino acids, are yielding to study and experiment in hundreds of laboratories. The intricate mechanisms of heredity, studied and found surprisingly complex in the simplest organisms, are becoming understood. These and other basic searches are not only progressing by themselves; they are also drawing closer together as new links in the form of common materials or processes are gradually uncovered. From all of this is emerging a better understanding of life. We are still merely on the shore of a great sea, learning the primary rules with a vast unknown before us. But what has been academic and detached until now is approaching the status of a body of proved doctrine out of which applications will flow. Some of these will again increase the power of man to destroy. So the threat of subversive war is real, and we may as well face up to it. At the same time, however, let us look at the other side of the shield, for there are always two sides, and we certainly cannot afford to look only at the power of new methods to destroy.

Subversive warfare and the opportunity for undercover attack of all sorts are no new things in the world. Yet the subversive has always been a secondary or auxiliary phase of warfare. This is not to say that it always will be, but there must certainly have been some reasons why it did not appear as a major phase of warfare in the past, and these may still apply. During the Renaissance, poisoning was developed to a fine art and was employed with gusto in palace feuds, yet there was no more than the usual minor poisoning of wells when one city finally came to open blows with another. During the old sieges, when a beleaguered town was being reduced to starvation and disease was rampant inside the walls, it was occasionally thought to be smart to load dead horses into the catapults and toss them in, but the military leader who did so seldom boasted of the fact, and it did not become general practice. The engines of war were still made to heave stones and bolts, or perhaps burning pitch, even when more deadly loads were readily available.

Coming down to the past century, it was common, in wars before World War I, for armies to lose many more men by disease than by battle. Then there were at least the beginnings of attempts at sanitation and disease control as a part of the military effort, the beginnings of the defensive application of biological science to war. During the contest just past, this deliberate application mounted to real proportions, and great things were accomplished medically; yet there was no corresponding effort aimed at carrying disease to the enemy camp. World War I saw the introduction of poison gas, but it was not used in World War II. In this last war every belligerent studied to some extent offensive biological-warfare methods, with the emphasis on being ready for defense, and much progress was inevitably made in offensive possibilities. Yet the only field use was sporadic and apparently due to uncontrolled military leaders. No nation put on a planned offensive by this method or even set in motion the preparations for doing so. The simple fact is that biological methods of offense in warfare have always been possible, and they have never been used in an all-out manner.

One can explain individual incidents by studying the conditions at the time. For example, the Nazis did not have supplies of advanced war gases until they had lost control of the air, and though their gases were superior to ours they did not know it. Similarly, the benefit we might have obtained by introducing gas against Japanese islands would have been offset by our vulnerability to the blister gases in a tropic setting, and, besides, we had many other ways of overcoming Japan once our full power was released and brought to bear. But the long history of the failure to resort to biological warfare is not to be explained by accounting for individual incidents, as is to some extent the use or nonuse of chemical warfare.

Accepted and traditional limitation of means in warfare certainly exists. It probably came to its height at the time of Frederick the Great, when all warring was done by professional soldiers, and most of their effort was in marching. One side quit and gave up a province as soon as it was maneuvered into a vulnerable position, so that overt clashes were in the nature of unplanned incidents. Something similar occurred in the Middle Ages. Footmen or peasants could battle mounted knights, as they proved very well at Crécy when they were allowed to, but ordinarily they were kept in the background, or slaughtered more or less casually if opportunity offered. The really decisive fighting was done by plumed horsemen incased in sheet iron, and even this fighting had its own set of ground rules. Much of this constraint disappeared with the French Revolution, when the people confronted the old order, intent on suppressing it, and the entire nation took to arms. It was the beginning, or rather a recrudescence, of total war, for the old wars of extermination were decidedly total. But not all tradition or limitation went out when total war came in, as the history of the World Wars indicates.

It is sometimes claimed that the limitation used to be a moral one, or more often one enforced by organized religion, that the morals of mankind have deteriorated and religion has lost its hold, and that now we may expect anything. It appears that we can leave organized religion out of the argument, for the wars and massacres carried out when organized religion, Christian, Moslem, or Jewish, was in control have not differed greatly in ferocity or depravity from those that were not thus controlled. But what about man's morals? Certainly we would not claim that there has been a great advance in mass morality in historic times; the extermination pens of the Nazis are altogether too recent for such comfort. But has there been a great decline? Or, to take the usual line of argument, was there a considerable improvement in the moral basis on which men waged war until modern times, and then a great fall back into barbarism?

It is doubtful if there has been much change. We can find organized brutality and debauchery throughout the history of war wherever we look, and in spite of Dachau it does not seem to be on the increase. On the other hand, if anything, the bright spots seem to be more prevalent than they used to be. If you wish, explain the freeing of the liberated Philippines, the decent government given to the people of Japan, or many another bright spot as enlightened self-interest; the spots are still bright. We know that if any of the great democracies conquered a people it would not exterminate them. It might have visions of profitable trade or buffers as it again set them on their feet, but we know it would set them up and not grind them forever in the mud of defeat. The morals of man change slowly. Perhaps we have no real reason for believing that they have improved since records were kept, perhaps conditions alone have dictated the ups and downs, but we certainly have no reason for believing that the race has suddenly lost all qualms, that no act, no matter how repulsive, is now barred if it will advance national interest, and that we may therefore expect any kind of dirty trick whatever.

If morals and tradition have little to do with it, then only the question of advantage will determine what methods of war are used. By this standard, if biological methods were more effective than others, they would be used. The case then becomes a matter of comparison, and we have to conclude that the reason they have not been used throughout the past is that they have never caught up in effectiveness with other methods. If, after reaching this conclusion, we predict that biological warfare is an active prospect for the future, we have to assume either that conditions have changed in such manner as especially to favor it or else that biological methods have made more rapid advance than physical ones. The second assumption would be hard to sustain. Though the biological sciences and their applications have lately gone ahead rapidly, so have the physical sciences, as proximity fuzes, rockets, guided bombs, and atomic energy amply prove. If biological methods are now to come in merely because they have made relatively greater advances recently, the case is not proved, and merely more or worse poisons certainly do not prove it. The first assumption, that new conditions favor biological warfare, we shall examine later.

But before that is done, we should emphasize that the question cannot be effectively answered simply on the basis of judged advantage; there is more to it than that. Without a shadow of a doubt there is something in man's make-up that causes him to hesitate when at the point of bringing war to his enemy by poisoning him or his cattle and crops or spreading disease. Even Hitler drew back from this. Whether it is because of some old taboo ingrained into the fiber of the race in its early days, whether it is merely an unconscious shrinking from a means that might backfire, it is there. It is also undoubtedly common to all the races of men, and it is very deep-seated. A review of the history of the mechanics of warfare indicates that biological methods have not been used because in comparison with other weapons they have appeared to be less effective. It also indicates that warriors have not even investigated and experimented to find out whether this was the case in practical struggle; they have simply refrained.

The state of mind of the scientists during the last war is decidedly pertinent in this connection. Almost without exception the scientists and engineers of this country rallied to the war effort. There were a few conscientious objectors, but no greater fraction than in the population as a whole, and an almost negligible fraction, as we all know. Among those who started on war work—and this includes nearly all the able scientists and engineers, young and old, who could find opportunity to do so— there was only one man in the ranks of the Office of Scientific Research and Development who quit because his conscience would not allow him to proceed on the development of means for killing his fellow men. He was a fine chap, respected then and now by his friends, but he was not joined in numbers. Scientists worked both on war weapons for carrying the fight more effectively to the enemy and on war medicine for mitigating the horrors of war, and with much the same intensity and skill in both. There was no enthusiasm, in the sense of pride in the overall accomplishment; the country as a whole and those who went into the armed services to fight directly did their work without the hoop-la of the first war, grimly, reluctantly, and with the sense of a disagreeable, tough job to be done well and finished with. But the devotion and performance were magnificent.

What did the youngsters, who could be expected to look for opportunity, most readily rush into, for they were often free in their choice of field of effort? When radar showed that it could determine the outcome of battles they rushed into that. They understood the objectives and possibilities of the atomic bomb, make no mistake, and they did not hesitate to deliver their best efforts to ensure its success. After all, when thousands of tons of high explosive were dropping on England there was no holding back on the effort to concentrate equal explosive energy in a single package for ultimate countereffort. There were qualms, of course—an intelligent man does not slay without soul-searching—but there was practically no scientist or engineer who did not take facts as they stood and strive to add his mite to advance the cause of his country and end the harsh task.

Perhaps the most repelling area of effort was incendiaries. The new incendiaries, dropped from aircraft, were the product of applied science and far more deadly than the somewhat crude forms with which we started the war. They were terrible things that spouted a sticky, fiery paste and viciously burned whatever men or materials they reached. The fire raids on Tokyo were far more destructive of the lives of innocent people than were the atomic bombs on Hiroshima and Nagasaki. Moreover, while they were exceedingly important in knocking out the war potential of Japan, they did not offer the saving advantage of the atomic bomb—the spectacular final blow that could end a war suddenly and thus save countless more lives than were destroyed. The people of this country as a whole took the use of fire raids in their stride; either they did not understand or, understanding, they did not object. It is notable that even after the war, when the whole affair could be viewed more soberly in retrospect, the few protests that were heard were directed against the atomic bomb and not against the fire raids, which were more terrible. There was evidently more to the protest than just a belated humanitarian impulse; fear was there, and fear for the future more than regret for the past.

When the fire bombs were being developed, how about the scientists? They worked on these in the same manner as on anything else and as effectively. Actually, in this area they had one of their hardest struggles with entrenched prejudice and obstinacy

among reactionary servicemen, a phenomenon that is as old as
the art of war, now on the way out, and not quite gone in the
last war. The scientists understood the terrors of incendiaries.
They were not lacking in human impulses. But if a war was on,
and incendiaries were being used, they did not hesitate to make
better ones; and when it was evident that mass use of incendiaries was necessary to get at the industrial production of Japan,
spread as it was throughout her frail cities, they did not protest.

How about biological warfare? This was under development,
if not in use, under a policy that sought defensive means and
studied the offense as a necessary basis on which to examine
defense. Moreover, the claims one hears of the effectiveness of
biological warfare are not a postwar development. There were
those, before and during the war, who fully believed that it
rivaled the atomic bomb and that, with an equivalent fraction
of the national effort devoted to it, its potentialities were even
greater as a means for bringing Germany or Japan to her knees.

Did the scientists rush into it? Did they insist that it be given
the attention its potential importance deserved? Did biological
laboratories all over the country turn their efforts automatically
in this direction? They did not. Were there energetic groups who
persisted in going ahead in spite of all obstacles, physical or
human, in the fashion probably best exemplified in the application to warfare of the physical sciences by the determined
group that produced the frangible bullet? Not at all. Devoted,
patriotic, courageous individuals reluctantly turned their efforts
in this direction in the laboratory and in offices because of a
conviction that we could not safely remain in ignorance of the
methods involved, and they did effective work. The medical
men would have none of it. Neither the Office of Scientific Research and Development nor the War or Navy Department
wanted it included in their organizational structures, and it was
tucked off in a corner in the maze of Washington. The National
Academy of Sciences advised on it, ably and wisely as is its
practice, at the call of the Secretary of War. But there was no
drive behind it. There was a reason for all this, and a deep one,

How is it today? If this is potentially one of the most powerful weapons of the future, great military thought must now be devoted to it. It is not. Do military men in the upper echelons study the fundamentals of biology assiduously? They do not study any part of biology at all. In the Military Establishment work on biological weapons is tucked off into a corner, under the Chemical Warfare Service. The total amount of money being spent on it, in all aspects, is a decidedly small percentage of the aggregate being spent currently for military research and development by the armed services alone, leaving out the large amounts going into further development of the atomic bomb. Devoted, able officers and scientists are working at it, in small numbers, but there is certainly no rush to join them. The laboratories concerned with it have more than the usual difficulty in securing a staff of able, trained men.

Should we conclude from all this that the possibilities of biological methods of waging war are a myth? It would be comforting to draw this conclusion, but it would be incorrect; they certainly have future possibilities of destruction of no small moment, even if in all soberness we include them as just one more possible system of weapons and not as an overwhelming type that would make all other means obsolete. They would be important in a military sense if they came into use, even if we discount the wild imaginings of those who contend they would wipe out the population at one stroke. Should we conclude that all military men are always blind to the possibilities of future weapons? If we were to do so in this case we should have to include the body of scientists of the country as well, for they were and are equally involved. It is simpler than this, and at the same time much more subtle. The human race shrinks and draws back when the subject is broached. It always has, and it probably always will.

There is a story that if one places two scorpions in a box they will fight, but they will not use their stings on one another, although the sting is lethal to a scorpion. It does not seem possible to find anyone who has tried the experiment, and the story may be apocryphal. There is also an account that when a hive of bees has two queens present, and should have only one for the good of the hive, the queens are brought to fight until one slays the other; but that if the queens arrive at a position from which each could sting the other simultaneously, they draw back. It is much more likely that these stories express something of a conviction in the teller rather than any absolute truth. Yet they illustrate the point; somewhere deep in the race there is an ancient motivation that makes men draw back when a means of warfare of this sort is proposed. It may be based merely on the fear that the means might indeed get out of hand and destroy the aggressor as well as his victim, or it may be more fundamental than that. In any case, it is there, and it has to be taken into account as we look forward.

The question whether conditions have come to favor biological warfare involves problems of general sanitation, the advance of public-health controls, the management of herds and crops, and the progress of medical science generally. The terrors of the possibilities of biological science, turned loose for destructive purposes and guided by skilled minds, are very real; yet they are less real when the level of living of a people is high, and when their guardianship of their possessions and themselves against the ravages of nature is on a sound scientific basis. We need to remember, as we examine what may possibly come, that all advance of science has two sides, that knowledge itself is neither good nor evil, and that the applications can not only increase man's power to disrupt and destroy, but can also protect. The last war was a terrible one, for whole populations were involved, and the destruction was imposed upon innocents and noncombatants, women and children, upon those who would have stood aside as well as those who fought directly; yet the new methods of saving life advanced rapidly along with new methods of destroying it.

For all its horror there were bright spots in the last war, and the brightest was the record of the medical men and those in the allied sciences who supported them. With sulfa drugs, penicillin, blood plasma, and advanced surgery, the mortality among the wounded was brought so low that the chances of survival of the wounded man who reached a front-line dressing station were extraordinarily high; some ninety-seven per cent survived of those who thus arrived. Whole diseases, notably tropical diseases, were overcome; the great scourge of malaria, for example, is a scourge no longer if we use the means to combat it that are now available.

An example of the power of wartime applied research to help mankind is given by cortisone, earlier named Compound E, an adrenal cortical hormone, prepared by a very complicated process from a substance found in beef bile. It has brought dramatic relief in experimental tests to those who suffer from rheumatoid arthritis, and promises great things in other fields as well. Recent attempts to synthesize it have been successful, and if its cost of production can be brought down it offers hope to many who are in distress and may well prove to be the greatest medical advance of a striking decade. This development emerged from one of the medical war projects. From preceding work we knew that the adrenal cortical hormones were essential to life and had powerful effects, some of which were bound to be related to the stress of war. It looked highly unpromising for a long time: this was one of the few projects on which the Committee on Medical Research and I did not see eye to eye. Fortunately they were right and we went ahead, for it is worth in the end, and in the light of postwar work, far more than was spent on the entire medical war research program.

Without doubt, the advances in medicine during the war saved and prolonged lives, and will continue to do so, to such an extent that the war as a whole, including its stimulation of biological and medical science, increased in the aggregate the chance of the young for survival to old age rather than merely cut down youth in its prime. The point of this is that the biological sciences, in their spectacular advance, are already producing marked results on the side of protection, and we have clear evidence from the recent past of their power in this regard. The theory that world conditions have so altered as to make biological warfare more effective, more advantageous as a weapon, is not supported by such facts.

With this background, let us examine the three forms in which the threat of biological warfare exists, considering now only its subversive aspects. The first possibility is that biological warfare might be applied as an extension of cold-war techniques, not to us but to free peoples still in a state of war exhaustion and hence susceptible, to bring them more easily under control when weakened by already familiar methods. Such action is highly unlikely. Unless the pressure techniques of cold war stop short of provoking open world war, they have misfired. There is, of course, always the danger of miscalculation, as Hitler miscalculated, but there is not much chance that methods will be used that would almost surely lead to immediate war—not in the present state of power unbalance.

We would certainly go to war at once if we learned that a potential enemy was slyly poisoning our wells. It is also pretty certain that we would go to war if scurvy tricks of the sort began to be played on other free peoples, for we learned the lesson of Munich, at great cost, but firmly. Any sporadic attempt at biological attack could hardly be decisive, and it would stiffen resistance and determination as nothing else could. The risk involved in making such an attack would be too great; it is highly improbable that it will be tried.

Second, there is the auxiliary use of methods of this general sort as a subversive accompaniment to total open war. There was a considerable amount of subversive activity in the last war, not by biological methods, it is true, but by other highly ingenious schemes. Some of them had their ludicrous side, as when explosives for demolition were smuggled into China disguised as pancake flour, and people even made pancakes out of the stuff, not good pancakes but at least without disastrous consequences. Nearly all of this sort of activity occurred in conquered countries, in collaboration with the underground. Some of it had real importance in annoying the invaders in various diabolical ways. Among the most thoroughly annoyed were probably Japanese officers who were sprinkled by Chinese youngsters with odoriferous sprays that would make a skunk blush. Some of the more overt activities were really important—for example, the very considerable disruption of Nazi rail traffic by the French *maquis*.

But it was always a side issue, which never fully determined the course of events. Even the Norwegians, bitter at their betrayal, unlimited in courage and determination, and supplied with some of the tools of their nocturnal trade, could make the Nazi conqueror uncomfortable, and undoubtedly derive great satisfaction from so doing, but they could not drive him out by subversive warfare or weaken his hold on their essential facilities. There was almost none of it in either Great Britain or the United States. The absence of real subversive effort was in fact surprising; even conventional sabotage for stopping the wheels of industry was almost nonexistent. No such events occurred as the Black Tom explosion of the first war. Part of the credit belongs to the keen eyes of the FBI. Still, there was more than this; the great democracies had become less vulnerable to that particular sort of mischief. Better organization, a population more alert because of civil defense programs, the fear of bombs and their plentiful receipt as far as England was concerned, with the consequent stiffening of all the civil machinery, made the field unproductive. After the war had progressed for some time, subversive warfare apparently was not even tried against us.

Would it be different in another war? No doubt there are more deadly tools to use than there were in the Second World War. But cattle diseases such as hoof-and-mouth disease or rinderpest, and the vulnerability of this country to their propagation, were well known before even the first war, and there was nothing mysterious or very difficult about the means for subversive dissemination of such diseases.

In the Second World War, we considered our enemy ruthless, with reason, for he exterminated millions of helpless people in cold blood, and barbarism has never gone further than that. Yet nothing subversive happened here. We should need to be more alert another time, undoubtedly; the kinds of things we should have to watch would be increased and the difficulties of detection intensified. Early infiltration of counteragents into subversive organizations would be at a premium. In peace we should certainly go forward with all speed on measures that pay out, anyway: expanded and improved public-health service, plant quarantine, research and control methods on plant and animal diseases, more highly protected water supplies, slum elimination, public education, furtherance of biological and medical research in all sound ways. If the threat causes us to do all these things more thoroughly, it will be all to the good. If war comes, subversive methods will play a part, they may even be disrupting to parts of our economy, but they will hardly be a determining feature of war unless there are technical changes that are not now in sight.

Finally, we need to consider subversive methods as an accompaniment to surprise attack on the outbreak of major war. Surprise attack itself by conventional means—another Pearl Harbor with fleets of aircraft and a difference perhaps only in the warheads—we have already considered. Here we are concerned with whether such a sudden blow by means of war that are now considered usual might be accompanied or soon followed by subversive destruction triggered off by enemy agents operating in our midst.

It might well be, and if the time approaches when the threat of sudden attack is imminent we should be on our guard in every way, with a civil defense organization in full career, external and internal intelligence agencies on their toes, and countermeasures ready. As the tension mounted we should need to guard ourselves in other ways. In particular we should have to keep close watch for enemy agents in our affairs, find them and toss them out. Could a democracy do all this?

We are not aggressively active about it at the moment, for we know that world affairs have not yet moved to this point.

But we can do it if we are keyed up to it, for we proved this during the recent war, and we can do it again even if the odds are a bit tougher the next time. Any such auxiliary to a surprise attack would require organization and planning on the part of the enemy, and a complete network of agents with communications and supply of materials. In spite of the insidious nature of biological materials, in spite of their potency in small doses and their ability to spread on their own, one does not knock out a great country by a detached sporadic effort. The organization would have to be perfect, the agents one hundred per cent obedient and circumspect. There might be traitors to the effort; there have been in the recent past, and an organized effort to poison a whole people or their fields and herds would certainly put a strain on any sort of indoctrination. One traitor, or even one careless agent, might spill the beans, and then under the intense search that would follow in an aroused democracy the whole thing might be given away. In fact, we may be more in danger of running wild on false scents than of overlooking real ones. A premature disclosure of plans for the auxiliary subversive effort might well wreck the main surprise attack itself. The risk to a potential enemy is large, and such results as he might attain may indeed be dubious. This is not at all to overlook the ravages that could be created on our cattle ranges if diseases were spread intentionally through them, perhaps by an innocent-looking plane dropping infected biscuits or the like. But even the shock of a virulent cattle disease completely out of control, even the virtual disappearance of our herds, would not bring this country to submission.

The greatest mistake the Japanese war lords ever made, and they made many, was when they attacked Pearl Harbor, for they thereby sealed their doom. We are a king in the countries of the world, and the old adage still holds that when you attack a king you must kill him. We should be alert by all means to the danger of surprise subversive attack; if we lapse into the sort of slumber that we indulged in during the twenties we shall indeed be in danger. But there is certainly nothing here to get panicky about. Subversive action is a dubious auxiliary to the sudden attack to open an all-out war. It might be crippling if things went badly. It can hardly be a substitute for war itself, for it can hardly bring a whole people to subjugation in the present state of the art.

The subject of subversive attack should not be dismissed, however, with consideration merely of biological warfare. There are other new methods, such as radiological warfare or the atomic bomb itself. These we have considered as a part of open war, but their possible part in subversive activity needs comment at least.

As far as radioactive materials are concerned, we can be brief. They are readily made in atomic piles in ample quantities for subversive use. If a little bit is placed in a man's desk, and he doesn't know it is there, it will damage him beyond repair, kill him if he stays around long enough. A little sent through the mail might put a recipient out of action if he kept it a while. The quantities necessary for attack through water supplies or through dust in the air are too great for ordinary sabotage methods. But the radioactive package left where it will do the most harm, killing those near by in a thoroughly insidious manner, looks like the real thing for subversive warfare of the thriller sort.

Two great disadvantages rule it out. First, it has no prejudice or predilection, and it will kill the enemy agent himself just as readily as his victim, unless he keeps it well wrapped up in lead or something equally heavy. Second, the radioactive material advertises its presence in no uncertain terms. A Geiger counter, that is, a properly designed black box, brought into its presence will rattle away in unmistakable fashion. It is easy enough, then, to examine mail or visitors or express packages to see whether radioactive material is present. If there is, the source should not be too hard to locate, should any have been left there, for the material advertises its presence wherever it is. Radioactive materials may well have their place in war, but it is not in the hands of subversive agents.

The atomic bomb itself is something else. The old scarehead

story that atomic bombs could be moved in suitcases may be dismissed. One does not lug atomic bombs about in that fashion. The idea of bringing in the pieces and assembling them in some innocent-looking warehouse or consulate is probably also far-fetched. It could be done, of course, even though a great deal of skill in assembly is called for. But one bomb thus planted would hardly be an effective opening blow in a war; and the simultaneous smuggling in of a number would involve great risk of detection, with the danger that one would be detected and the rest seized, to be returned to the source with interest. Moreover, these things are expensive, as we have discovered in this country, and they would not be risked in many readily recognizable pieces if there were other and surer ways of delivering the finished article.

Unfortunately, there are. We have considered delivery by aircraft or submarine in discussing open war. But there is another method that could be a decided peril in subversive warfare. This is to carry atomic bombs in innocent-looking merchant craft that can be moored near cities, to be blown up bodily when the time for surprise attack comes, with the crews presumably scuttling off just before if their home government is benevolent enough, and trusting enough, to give them warning. More flexible, and hence more to be feared, is the scheme of planting the bombs on the bottom of harbors, or in canals, through trap-doors in the holds of merchant ships, to be detonated later by time fuze, or even by radio signal at the expense of a considerable amount of gadgetry and some risk of failure. Presumably, bombs thus placed near various cities would be detonated all at once, as the opening shot in a total war. There is no trick method of determining from a distance—or close at hand, for that matter—whether a ship carries such a deadly load. The atomic bomb does not advertise its presence in the unmistakable manner of highly radioactive materials. In fact, an inspection would have to be very thorough to find one, for it would be disguised in an innocent-looking case or buried under conventional cargo that could not readily be moved. The damage to a city could be very large if such an infernal machine went off in its harbor, not so large perhaps as some of the accounts would indicate offhand, for there are limiting factors, but large enough. We have the experience of Bikini to guide us here; the wall of radioactive mist that spread from the underwater test was a very terrible thing.

The limitations are not very comforting, but they should be mentioned. There are not many harbors with sufficient depth of water for full effect, and such places are not ordinarily at anchorages. Much depends upon the direction of the wind. If the instant can be chosen—that is, if remote control of the bomb is to be employed or the carrying ship is itself to be blown up— the proper instant in view of wind direction can be chosen. It will hardly occur in many harbors at the same time, and simultaneous detonation is necessary; otherwise every suspicious ship will be promptly seized or moved out and the bottom in suspicious locations explored when the first explosion occurs.

But even one such bomb can make a considerable fraction of a large city uninhabitable and useless. It will not necessarily kill enormous numbers of people. Those close to the detonation will of course be killed, but the large numbers in the area of drift of mist, if they move out promptly, need not be killed. Of course, such a movement itself may well cause panic and great loss. The area affected cannot be used, probably not for a very long time. Decontamination methods are difficult and generally unsatisfactory. This is not at all the same problem as with a poison gas, which will drift away if vaporous, and can be altered chemically if a liquid spray, and thus rendered innocuous. It is not even the same as biological toxins or bacteriological agents, which can be countered, at least if we get at them chemically. For the radiation products of the bomb, spread in the wall of mist, have the serious property that they remain equally damaging no matter how much they are altered chemically. Undoubtedly decontamination of a sort could be carried out eventually; and the radiation products decay, some rapidly and some very slowly, but with the more active ones disappearing first. Yet by this subterfuge a whole section of a city could be made uninhab-itable for a long time, certainly enough to take it out of the war effort, and such a loss could be highly disrupting. With our many coastal cities, we are particularly vulnerable. We would not cease to fight a desperate war even if many of them were made useless, but the blow would be a heavy one.

Now, of course, if tension mounted, in the days when a poten-tial enemy had the atomic bombs to use in this way, we should quit welcoming his ships and shepherding them into select places near our cities. We should probably quit receiving them at all. We should not need to be quite so careful with ships of his satel-lites or potential allies, for there is not much chance he would place one of his bombs in the hands of doubtful allies; doing so would involve rather a large chance that it would arrive at the wrong destination. The danger is that the tension and the warn-ing would not be there, that the stroke would come out of the blue when relations were apparently smooth. We just cannot afford to be gullible.

These, then, are the main forms that subversive warfare might take. They are not pleasant, no war is pleasant, and one of the most unpleasant phases is that of the stab in the back, or the terrorist who is our guest. But, on the other hand, they are not terrifying, any more than all of human existence is terrifying while men still fight. We can cope with them, if we are not lulled to sleep by the blandishments of those who would destroy us or by our own self-delusions and laziness, and if we are not hyp-notized by abandonment to unreasoning fear.

The way in which we counter the threat of subversive war is clear; we counter it in the same way that we prepare for total war, by building a strong democracy in every respect. We fur-ther the biological sciences in their generality, not merely in their possible wartime applications, see that there are attractive fields of effort so that young men of great intellectual capacity will enter them in numbers, and see that there is no financial bar to their doing so when they have exceptional ability. We perfect our medical and public-health facilities, our sanitation and our hospitals; we advance the medical profession vigorously, making sure that in so doing we make it increasingly possible for men of independence and mental grasp to operate in them without the deadening hand of bureaucracy over them. We make our governmental machinery more streamlined and less permeated by confusion, so that if there is a task to perform, such as the perfection of methods and organization for handling suspicious ships, at least it will be known whose job it is, and so that we shall have a chance to proceed without impediment by political shenanigans. We build our organizations, governmental and private, labor and industrial, charitable and quasi-political, so that we are less vulnerable than now to the wiles of the traitor or enemy agent who would bore from within. We make our system work, the system with which we have already gone far, the democratic system, without abandoning in the process the essential human freedoms upon which it is founded. We build our strength, and face the future, atomic bombs and biological warfare and all, with the determination of free men whose destiny is in their own hands.

CHAPTER XII

COLD WAR

"Cold wars cannot be conducted by hotheads. Nor can ideological conflicts be won as crusades or concluded by unconditional surrender." —WALTER LIPPMANN
The Russian-American War. 1949

THIS IS A BOOK on science and democracy and war and their relations to each other. In these days cold war is an important phase of warfare, and we should not leave it out. It is primarily a political rather than a scientific affair, and therefore somewhat off our path. We shall treat it briefly, however, for it does influence our approach to those aspects of war in which the operation of science in a democracy is the central element.

How do we counter the tactics of cold war? Fighting such a war is a complicated matter, and we are none too good at it as yet, but we are learning. If we are clever at a bluffing game—and we have always rather prided ourselves that we are not too bad at poker—the task ought to be along the lines of our national aptitudes. But committees and public discussions are not conducive to skill in a bluffing game, no group conferences were ever a successful adjunct to playing a hand of poker, and we have a feeling of disadvantage in a contest with a tight, secretive, hard-bitten lot of players on the other side. Besides, though we like poker, we don't like this game.

To pin down what we mean by cold war, let us consider the two divergent concepts of the relations of nations during peace. One, and the one we should much like to adhere to, is that their normal relationships are those of peaceful trade, that wars are accidents brought on by avoidable clashes of interests, or planned aggressions by buccaneers who could and should be suppressed. We should like to think of peace as the normal condition, and war alarms as a vestige of a sorry past that we should soon out-grow.

The other concept is not like that. It considers war as normal, and merely the extension into active fighting of an inevitable contest among rival nations, some of which are to rise and others to fall. Peace is merely a breathing space, an opportunity to recover war potential, and an opportunity also to work aggressively by all possible means for the weakening of the obvious enemy—winning over his possible allies, obstructing his trade, and confusing his beliefs.

There will never be a world government, or even an agreement among nations capable of securely maintaining peace, until a sufficiently powerful part of the peoples and the nations of the world agree that the time has come to adopt the first philosophy and contain those who hold to the second until they, too, see the light. We have not arrived there yet, but the chances of arrival are probably better than they have ever been in the world's history.

In the meantime we have to contend with those who regard peace as merely an interlude during which methods short of war can be utilized to seize territory, conquer peoples, and spread confusion and internal strife among us. This is cold war. It was not invented by Hitler, but he brought it into its modern form. Its greatest tool is the threat of war and the spreading of fear to weaken resistance to external demands. Hitler carried it out so successfully that, one by one, he took over many of the facilities he needed to build his strength. He had no intention of departing from this highly successful method, but he slipped, by reason of the resolution of England and France at the time he invaded Poland, and plunged the world into open war, though only after he had so enhanced his power by cold-war tactics that defeating him became a desperate undertaking.

Again the method is in full use. But there is a difference this time, and it is an insidious one. Hitler, when he wished to take over Austria or Czechoslovakia, was limited to the direct threat of armed invasion, propaganda to terrorize and weaken, and

minority insurrections within the victim country by which he
finally seized power. We have seen all of these used again, but
now they are joined by another force.

This is the Communist thesis that if enough distress and mis-
trust can be stirred up in a country, its working people will
revolt, a Communist regime will inevitably follow, and the country will automatically became a new unit in the Communist
system of totalitarianism. This is the greatest threat, for it is far
easier to tear down than to build up, far easier to spread lies
and turn men against one another than to establish mutual confidence, far easier to wreck a political system than to cause it to
operate smoothly. This is the toughest sort of cold-war tactics
to counter.

There is little doubt about what we have to do to meet part
of the cold-war technique. First, we have to confront force with
force on frontiers. Our presence may be a token, but it is an
exceedingly important one. It is well known that we rouse
slowly, and hence that a seizure of power in a neighboring country can be made quickly before we move, confronting us with
an accomplished situation, and that we will not then go to war
to redress it unless committed in advance or unless a succession of such events comes too closely together. But it would be
another thing to attack and overrun our troops. There is no doubt
whatever that we would then strike, and that we would plunge
in completely, and fight to a finish. We have demonstrated the
way we carry on war. Hence our troops will not be overrun at
present. All sorts of methods will be used, but not that, unless
we are exceedingly foolish and make it inevitable by placing
them in impossible places. Also it has become clear that, if we
are fully committed in advance to resist an explicit conquest,
it will probably not occur. But our commitments in this respect
have to be definite and wise; we cannot afford to overreach ourselves. Neither can we afford to let vagueness imperil those on
whom we would depend if the cold war should lead to all-out
conflict. Not all this is fully clear in the minds of the people of

this country, but understanding of it is spreading, and there is growing public support for a firm stand.

Second, we need to help our friends, and we are doing so. We are taxing ourselves to send food and machinery to Europe, and we shall continue. This is not charity, and it is not so regarded. It is enlightened self-interest; we are spending money to strengthen the part of the world with which we have common cause. It is an expensive process, and it is straining our resources to the limit and risking our entire economy. We could overdo it and force inflation. But this is a rugged country, economically and industrially, and can stand a lot. In any case, the aid we are rendering is the best investment we could possibly make in the direction of preventing war. As we join direct military aid to our economic aid, we make the best possible military investment. A dollar will go far, when costs are low, if it is intelligently placed, and our object is to attain the maximum strength for ourselves and our friends.

Some opinion holds that we are overdoing this matter of preparedness and aid because we are gullible. The argument runs this way: The Communist regime that we face knows well there is to be no all-out war in the near future. It has no intention of starting one or of allowing incidents to proceed too far and bring one on, for it knows well it would be defeated after a struggle. That regime also knows, for all it tells its own people the contrary for its own purposes, that the great democracies will not start a war unless under enormous and direct provocation. The Communist powers take no stock in the tale of our starting a preventive or prophylactic war. (Neither should we, for democracies are not built that way.) On the other hand, they believe fully that the capitalist system will crack, that it inevitably involves booms followed by severe depressions, that it will get itself all snarled up, and that this catastrophe will be their chance. When we are thoroughly in industrial and economic confusion, when we can no longer carry great national budgets without deficits that would wreck the stability of the dollar, then they can safely make the next conquest and move ahead. For this reason, the argument goes, they behave as abusively and disagreeably as possible, so that we will overreach ourselves, create great war machines that will never be used, exhaust ourselves in so doing, and hasten the inevitable collapse.

Now, there is not the slightest doubt that we could thus over-extend ourselves, and the corollary is that our military expenditures and our expenditures for aid to our friends must be made within reason, and with careful logic, correlation, and economy. There is also no doubt that the most important thing for us to do, to maintain our full strength and bring the world back to sanity, is to keep our industrial and economic health, keep the machine running at full blast without inflation or depression. Should we fail in so doing, the resulting distress in the world would play directly into the hands of those who would build on chaos.

There is, nevertheless, a fallacy in the argument that the strident belligerence, the intentional abuse and vituperation, to which we are treated daily by the Communist press are more than propaganda for spurring internal action, or attempts merely to play upon our fears, and have been deliberately planned by master minds to lure us on more rapidly to a crash. The master-mind fiction will not stand up in any case; the so-called master mind makes too many obvious mistakes. But the fallacy is further evident in the results. Under the spur of our conviction that we still live in a hazardous world, under the stimulation also of the shrill, parrotlike repetition of insult and challenges, we have started to rearm. We have also started to help Europe back to strength. And the Politburo certainly did not intentionally goad us to that action, for a prosperous and rearmed Europe, having as many people in it as all of the Soviet Union, containing the best skills and most developed resources of the Continent, would be in and of itself an almost insurmountable obstacle to world conquest by armed might, an almost complete barrier to infiltration and intrigue, provided only that it be united in intentions and stable in its political institutions. Moreover, in concert with this country, it could without the shadow of a doubt maintain the peace of the world by armed might. So, if "planning" in Moscow contributed to this action of ours, it was colossally poor planning.

It is clear how we can proceed to counter the principal methods of cold war. We maintain our full military strength and place our forces where they defend our vital interests and where it is essential that there be no advance against us. We aid free peoples to recover and become strong, economically and from a military standpoint. We keep this purpose clear. Our funds and our aid are not being expended to increase the area over which we hold sway. We do not need to argue that point; the free world knows it. Neither are they being expended in a crusade to spread our economic beliefs or our particular system of free government, although this point does need emphasis, for there is occasional confusion about it. Nor are they being spent just to relieve distress. That is a necessary and highly desirable accompaniment, for distress must be mitigated before there can be hope and courage, and these are essential to recovery and strength. But our aid is being expended to produce strength in the free part of the world, among potentially powerful nations that we feel sure share our desire for peace and will remain free and independent. Hence it should be placed in the manner that will bring the greatest strength for the expenditures we find possible. It is reasonable to insist that it be used only where it will genuinely strengthen and not be diverted to other things. The program has been well handled and effective thus far and can so continue.

As we meet cold war in this way, time can work for us. The men in the Kremlin—probably today with some dissenters who begin to doubt—apparently think differently. They believe time is on their side because, as they see it, their system is economically stable and ours is not, and therefore all-out war is not probable in the near future. They base their beliefs on the conviction that the capitalist system will break, that it contains the seeds of its own destruction, and they wait for the event. Looking back over the history of the world since the industrial revolution, they see a series of booms and depressions, overproduction and underconsumption, suddenly creating a downward spiral that plunges nations into turmoil, and they expect this to be repeated on an enlarged scale.

It is a challenge we must meet, and to meet it will require our best thought and patriotism. We do not need to cancel all the ups and downs of business—some of this fluctuation is inevitable—but we do need to prevent the downswing from suddenly cutting off the entire livelihood of great numbers of people, and we do need to accomplish this without permanently weakening the national credit. It is equally essential that we prevent the upswing from driving us into inflation, or into the crazy boom of the Florida headache, or the absurd stock market of 1929. In times of prosperity, by economies and restraint, we need to strengthen the national finances in every way, to be able to meet the days of adversity with latitude still available in the national credit.

It may be impossible to do much of this in the immediate postwar period, with heavy armament on our books in peacetime for the first time in our modern history and with the necessity of aiding our friends. But as this load lifts, as it should as the world recovers, we have the disagreeable job of curbing our enthusiasms and extravagances in the interests of stability. In times of prosperity, and as far as we can in adversity as well, we need to do more than this. We need by tax policy and other means to stimulate free and competitive enterprise, to build more modern plant and equipment so that we can produce more and more useful goods at favorable prices. We need to have industry, labor, the farmers, and government joining in action that not only will make the economy stable, but also will make it grow in a vital way. It is not yet proved that we can do all this, and the test is yet to come. But we certainly have the need for the accomplishment before us in no uncertain terms, underlined in every day's news, and the country as a whole seems to realize what it faces. That is something.

That time can be made to run in our favor in a military sense appeared from our examination of the probable nature of future war. Time can run for us indeed if we strengthen the organization of our military machine, carry on with vision and good sense every important aspect of the development of military devices, and at the same time maintain our economic system healthy and prosperous. In these ways we can meet the primary aspects of cold war as Hitler practiced it: the direct military threat on the frontier, the propaganda of lies, and the sowing of economic distrust. We can do so with the conviction that time runs in our favor.

It is the other aspect of cold war that is baffling, the boring from within, the utilization of the advantages of a free democracy to undermine it in countless ways, the penetration to positions of authority and the misuse of that authority for the objectives of a foreign power, the playing upon internal prejudices to further strife, the existence of a disciplined network of spies and saboteurs in our midst, taking its orders from Moscow.

It is not likely that other free peoples of Europe will be easy victims of such tactics from now on. It is certainly to the interest of those who remain free, and to us, to see that subversive seizure of power does not happen; and the support of legitimately constituted governments by neighbors, and by us, will be much more readily effected as the standard of living rises. The world seems to have seen enough of subversive politics and to be ready to meet it. It is for every people to determine how they will be governed, and we should not interfere, even if we disapprove. But the form should be determined by the free and enlightened will of the people of the country themselves, not from outside, and a seizure of power engineered by Moscow, on the pattern of Czechoslovakia, is not tolerable. It need not happen again if we stiffen our backs. With free peoples, who are evidently going to maintain their freedom and control their internal traitors, we can combine as well as aid. We have done so, in the Atlantic Pact, and this is a milestone on the road to peace, for the destinies of free peoples are becoming linked under specific agreements of mutual aid if attacked. We can aid them in a thousand ways, as they bring order and strength and stability out of the postwar turmoil, and resist the tactics of Communist infiltration in the process.

But we cannot do the same in China, and the gradual infiltration in disorganized countries, as in China, is now our most serious problem. Here we cannot draw a line, with our own troops, beyond which we will not tolerate aggression by any means, for the area behind us would be soft and malleable, and we might well be placing troops in an impossible position and asking for the very conflict we seek to avoid. Neither can we sit idly by and watch the seizure of all the weak areas of the earth, and their organization against us, for they contain hordes that could be trained for war and material resources that could support a formidable military machine.

What can we do, in such countries where there is overpopulation and underproduction, ignorance and distress, and political chaos? We certainly cannot simply watch these areas, with their enormous natural resources, be absorbed one by one into Communist totalitarianism by the familiar process of support of puppets and the sowing of confusion. Neither can we step in with armed force, alone or in concert with other democracies, without overextending ourselves. The people of those areas, we hope, will erect their own free systems, further education, and take their places among the independent democracies. So some of them will, and we would aid them to do so; but this will require a generation at least, and in the meantime they may be conquered by the new techniques masquerading in the name of liberty and aid to the common man, and then perhaps become integrated into a tight system whose avowed purpose is to bring the whole world into that framework, with all freedom suppressed.

Imperialism, in its old form, is ended. At its best it furnished stable government to great areas until they could become independently stable, and furthered their progress by protection of investments and trade during the dangerous interim. Our experience with the Philippines, for example, came out well, in spite of vicissitudes, but this sort of thing is over, except for Japan and Germany as conquered countries, where we hope it soon may be over. Imperialism of the less happy sort is also ended, where investment and military control marched together, not for the purpose of placing a people on their feet, but for exploiting them at low real wages and gaining access to their resources for utterly selfish purposes.

The problem of these areas has intimate technical aspects, for what these countries need is communication, roads, railroads, trucks, telephones, broadcasting, to bind them together. They need scientific agriculture to overcome their famines, great fertilizer plants, agricultural machinery. Above all, they need education, and this means a higher standard of living with consequent leisure. But we cannot just pour these things into disorganized areas, even if we have them to give. They will merely be seized and turned to other purposes by those who grasp military power and seize the reins of government. Then, when that government arises which maintains itself by alliance with Moscow, through the elaborate system of control Moscow is busily perfecting, we have lost.

That system is not infallible, of course. The Soviet scheme of managing satellites is cumbersome, and too intricate and widespread a dominion managed by that scheme might well fall apart in time; but we cannot count on that, even with the obvious stresses before us, as in the case of Tito. It is essentially a very old system. In the days of the Great Khan, when the Mongol Empire spread over most of Asia and much of Europe, the seal of the Khan was bowed to over a vast area, even though it took him three months to communicate with one of his outlying satraps. In these days of rapid travel and communication, with the world shrunken by the airplane and the electric wire, it would appear much easier for a dictator to extend and maintain sway over vast areas, especially with the aid of the tank and machine gun to control mobs. But there is an offset. The power of the ancient regimes resided in the ignorance of the masses,

excluded from foreign ideas, so that they knew and submitted
to only one system. Iron curtains are no new invention; yet they
are now much harder to maintain. The same technical advances
that sustain in mystery the distant emperor, whatever his modern name, also tend to penetrate the barriers to ideas that he
must maintain for his continued sway.

Here lies one of our opportunities. We certainly ought to be
able to spread ideas, if we have them to spread. If the amount
we spend in the competition between almost indistinguishable
cigarettes, or between flavored brands of alcohol known as whis-
kies, were used for the purpose, if the money spent in advertising
cosmetics were diverted to the purpose, we could tell a very
large number of people many things. They can be simple things,
but they are worth telling.

Yet none of us would wish to rely on this activity alone, even
if we put our strength behind it in genuine fashion. The power
of ideas is great, given time, but the time may be very long indeed when there is a force ready to exclude and distort them,
backed by a secret police with tommy guns. We need to further
understanding of our true way of life to the maximum extent
that the printing press and the radio wave can reach into the
hinterland, but that is not going to be enough, as the confused
position of nations in the world order becomes determined.

There is one more thing we can do. We can become a bit
broader in our outlook as to who may be our friends.

The free democracies of the world are uniting and guarantee-
ing one another support if their system should be attacked by
an attack on any one of them. Within the framework of the
United Nations, and looking forward to the day when that organization may grow to strength and genuinely impress peace, they
are coming together for joint military action if it is needed. As
they do so they will gather resolution to fend off the assault
from within by subversion and intrigue. They will further one
another's well-being by trade, and a lowering of artificial barriers, as relative prosperity returns to the world, as it is bound
to do in time.

It would be folly to jump from this to the conclusion that the world is to be divided into two camps, democratic and totalitarian, and that hence every country that does not have a democratic regime is to be forced into the Russian orbit. Democracy involves habits of restraint and tolerance and respect for other people's rights, reverences and understandings and deep effective emotions about these things, which are arrived at only slowly. There are countries where true democracy is impossible and will be for a long time to come. These will be ruled by dictators of one sort or another, especially by adroit cliques. Some will be merely jockeying for position, attempting to embarrass us, or anxious to give lip service for ulterior motives, and these we may as well allow to continue to stew in their own juice. But must we consistently exclude them all from collaboration of all sorts? When we were fighting Hitler and Mussolini, the totalitarianism of the right, we did not hesitate to ally ourselves with the totalitarianism of the so-called left. In fact, if it had not been for Russia, we should have had a tough fight indeed. We do not need to believe the Russian stories, for home consumption, of the superiority of their war performance and the ineptness of ours, to realize that this was a fortunate alliance. Now that we are faced with the threat of Communism we should not be so strait laced as to forgo every advantage in the struggle by becoming rigidly doctrinaire.

With the true democracies we would bind ourselves tightly. We will even aid them, by all legitimate means, and there are many, which do not involve an improper entry on our part into their internal affairs, to maintain the stability of their democratic institutions in the face of internal plots against them. We cannot, and should not, go so far with the country that is not democratic. But we can be friends with it, if it is bent on peace, and we can help to build it up in reasonable ways. Someday it may be democratic, we hope, and most dictators will at least join us in expressing the hope. When or whether it does are different questions. In the meantime we wish to trade with it. We especially wish it to recognize that there is a power in the world bent on peace, bent also on decency between nations, to which it is to its advantage to adhere.

There is room for plenty of realism in the world as it stands. As we combat Communism in a cold war we should not lose our ideals; we should preserve and even advertise them. But this is not inconsistent with good relations with the rulers of any country who do not or cannot for the moment subscribe to these ideals in detail, provided only that we can maintain honest and mutually advantageous relations with them. If the government of such a country seems stable and willing to give proper assurances that investments will not be unjustly seized, we can encourage our capital to enter it, to the extent that is safe. We have to do this in the light of the fact that we do not intend in any circumstances to try to rescue it by armed force, and the principal reliance must be that investments will be safe because the sacrifice of overall advantageous relations for the purpose of a small seizure would not make sense. We can arrange reciprocal trade relations with such a country, even if doing so makes one of our own industries squirm a bit at times, if reciprocity will actually advance our own prosperity somewhat and the other country's greatly. We can lend technicians and render their service dignified and attractive by proper recognition. We can even aid in military ways at times, on a *quid pro quo* basis, with information and the like given in return for commitments or even bases. We can tie such a country to our orbit, rather than to the Russian, by making the choice clear and adherence to us advantageous. We can make democracy and prosperity synonymous, and sell this realistically and vigorously. But in order to do so we have to make democracy work.

There are many facets to this difficult subject of cold war. If we have to counter its threat, by all means let us learn the methods and counter effectively. We look forward to the day when cold war will no longer be a feature of world relations. In the meantime it does not need to get us down, if we understand it.

CHAPTER XIII

THREAT AND BULWARK

"It is mathematical intuition that urges the mind on to follow the mirage of the absolute and so enriches the intellectual heritage of the race; but when further pursuit of the mirage would endanger this heritage, it is mathematical intuition that halts the mind in its flight, while it whispers slyly: 'How strangely the pursued resembles the pursuer!' "

—TOBIAS DANZIG

THE WORLD is split into two camps. In blunt summary terms, there are on the one hand those who believe in freedom and the dignity of man and on the other hand those who believe in a supreme conquering state to which all men would be slaves. After the close view we have taken of the instrumentalities that science offers to arm these two camps for a future clash—if one is to occur—it will be useful to step back to take a long-range view of the division. Often, putting matters in the starkest possible contrast, in idealized and extremely general terms, helps to give sound basis for later specific judgments.

There have always been two contrasting philosophies of the relations of men; there have always been aspirations toward freedom and systems for holding men in subjection. But the sharp division has now become crystallized and the lines are clearly drawn. The crystallization is sharper, the lines of division are clearer than ever before, largely because the scientific approach to life has entered into the common thinking of men and has there produced distortions. The future will be influenced by the ways in which science affects the material things of man's existence; it will also be determined by the way in which his interpretation or misinterpretation of the teachings of science affects his philosophy of life, by its effect on the choice before him. Mankind has come to a parting of the ways. There is now a fork in the road, and two courses lie ahead.

171

On the one hand lies a continuation of the past, a continuation of the rule of tooth and claw, of recurrent wars interspersed with periods of uneasy peace during which the struggle continues underground, of increase in the power to destroy through modern propaganda and modern weapons, of no refuge in defense or distance, and finally of a return to barbarism and small-scale struggle when destruction has again limited the power of the offense by breaking down the machine and the organization that make it operate. Along this path we are promised a world government created by conquest, holding enormous central force, and overturned at intervals as long as the spirit of freedom survives, until at last it takes a form that is ruthless and skilled enough to hold the full population in physical and mental thrall under an oligarchy that somehow learns to create a new and very objective system of self-perpetuation effective enough to control the ambitions of the elect.

Along the other fork of the road lies the world of self-governing peoples, creating health and prosperity in their midst, collaborating to maintain the peace and restrain the aggressor, until finally there emerges one world governed by a system truly responsive to the enlightened will of the whole people.

The world has long been torn by struggle over the choice between these courses; we did not come to its end when the last war ended; it will extend long into the future. The outcome will depend on faith, faith that there is more to a philosophy of life and more to the nature of man than harsh, selfish struggle for survival and domination.

To set this clash of philosophies, this struggle over the choice, in perspective for judgment, it is worth while to review in capsule fashion how we came to where we are. Many of the arguments of those who see only a harsh and materialistic outcome of the struggle are based on their interpretation of the nature of the path we have already trod. We shall trace this briefly and in the stark harshness that usually carries conviction to those who see for the future no choice at all, only a continuation of the road of the past, with an improbable path branching off that might be passable but that no self-proclaimed realist would regard seriously. We have to reckon with this view, for many who have held power in the world in the recent past—and who hold it now, for that matter—attempt to teach it to multitudes and grow in power because of the desperate conviction thus inculcated in their vassals that there is no better way. Moreover, it is often held that he who would attempt a logical outlook on the world, the scientist, for example, must be driven by his own logic to this view, and a statement to the contrary would perhaps be salutary. So this brief review of evolution will place its emphasis on the more crude and mechanistic aspect whenever there is still disagreement or a choice.

Some time in the past, a few billion years ago, there was an event. As a result, myriads of suns were sent spinning on their way in a three-dimensional cosmos, probably finite in dimensions but unbounded. Whether this event was a cataclysm that exploded a concentration and produced an expanding universe, or a gradual agglomeration of previously existing, sparsely spread material, does not matter for our purposes here; there was an event. It also does not matter whether or not this particular event is called the act of creation. No physicist or astronomer will attempt to step back of it in his accounts, nor will he contend that it is or ever will become revealed in its entirety to a finite mind. For if it did, he would promptly have to replace it by an earlier and more nearly fundamental event. It was a beginning, the nature of which is beyond human ken, and this is enough for our present account.

More pertinent to our story, there was a later event, when one of these whirling stars became disrupted, or condensed into more elaborate form than its neighbors. In fact, there were probably many such events, for stars were plentiful even if the opportunity was rare and even if only one such event touches our experience. It sent fragments of the star whirling by themselves in a system of planets. One of these, as it cooled, came to a rare condition where liquid water and an atmosphere existed and, rarer still, where carbon compounds in solution could ring

their multitude of changes and variations, absorbing and interchanging the energy received by radiation from the sun.

Then another great event occurred: a compound was formed
that could interrupt the complex dance of chance interchange
and chemical transformation, a compound with properties such
that its very existence caused more of it to form by some obscure
molding process—and reproduction appeared on the earth. Thus
was produced a one-way trend in an otherwise chaotic state,
for this new compound inevitably aggregated to itself all the
available material capable of being made to take the irreversible
step, or the step that reversed too slowly to offset the cumula-
tive power of the new molding process of forming replicas. Thus,
if we accept the current discussions, was life begun.

Now the opportunities widened, for any chance change that
produced a new compound with this power of aggregating to
itself by reproduction the surrounding complex materials could
proceed on its way. And there were many such, and they came
into competition for the materials available. Each success paved
the way for a later success, but one more complex in its composi-
tion. Thus appeared what we call organisms, competing for the
energy and carbon of the earth in its primordial seas.

The next event was profoundly important. Some pioneering
organism learned to prey on another, and to seize upon compounds in which the important initial energy transformations
had already been performed. This circumvented the need of the
elaborate primary mechanism to utilize sunlight, and conflict
came into the world with the beginnings of the animal kingdom.
Thereafter there could be no peace in a world governed by the
struggle for survival, and the possibility of complexity in evolu-
tionary forms became immense.   .

There was a premium on mobility. The organism that could
move could battle better than the one that could not, could compete to better advantage, could monopolize food and replace
sluggish competitors. So organisms learned to move about.

From then on the story is merely one of the elaboration of
complexity for advantage in the conflict. Concealment and es-

cape had their place in defense, but the great elaborations were in offensive means, the creation of new species capable of existing by preying on others, and built to fight in the battle for survival, or to seize new territory and defend it against attack.

Now the elaborations defy concise and orderly summary. Nerve systems evolved for the control of large and complex competing organisms. Temperature control yielded great advantages, and mammals appeared. Enormously complicated genetic systems and the whole gamut of the sex mechanisms yielded advantage in storing and combining mutations as ammunition for the variation of species necessary to follow changes in environment and to create new species to occupy new niches, and so they were preserved and elaborated. Senses were evolved by which the organism could apprise itself of its surroundings, and thus secure an advantage in conflict, by reason of electrical systems in the organism that responded to vibrations of light or sound, to pressure, or to chemical stimuli. Central nervous switchboards appeared for correlating these sensations with the optimal consequent acts and movements—the acts best fitted for escape or attack in the ever-present struggle. As relationships of an organism to its environment became more and more complex in these and other ways, the little knots of nerves gave way to a central brain, and this evolved to such extremes that the offspring were endowed with complex patterns of behavior already established in the inherited arrangement of nervous interconnections, so that the spider could spin a web without ever being taught, and the young eel could find its way back from the Sargasso Sea to the streams of its parents without a guide. Parasitism came in, with disease as one of its manifestations. Evolution, rushing blindly on and seizing every temporary advantage, made its long-term mistakes and entered dead ends with over-specialized forms incapable of further adaptation, such as the dinosaur. The intricate genetic mechanisms that were capable of applying chance mutations in combination to produce desirable modifications became adapted also to produce the grotesque. Life became elaborated into multitudinous forms, adapted to occupy every nook and corner of the varied world of nature, in one vast fight for existence.

Communication appeared between individuals, for warning or concerted action. Societies developed, as advantage accrued to the organism that learned that combinations of individuals could sometimes compete to better advantage, and these evolved into every kind of society from the nest of ants to the pack of wolves. There also appeared an advantage in the struggle for those species which prolonged youth and the teaching of youth, for supplementation of inherited capacity by direct guidance was a powerful aid in meeting the vicissitudes of existence. Thus entered into the world protection, care of the young, the maternal instinct, and these gave rise to a duality in which the gentler virtues took their place alongside the aggressive ones as assets in the struggle.

Just when memory and conscious thinking appeared may be disputed. But the dog dreams, and the fox pups play, and any trapper concedes substantial mental capacity to his victims. The great event occurred when these were joined by curiosity. The change appeared, perhaps, not so much because the simian developed hands in the trees, and found that the grasping of branches led readily to grasping a tool, as because he was curious about the result. At any rate, conscious evolution dawned, life embarked on a new adventure, and man appeared.

His stay has been short, compared to all that went before, but it has most certainly been hectic. Conscious cerebration, reasoning, even if elementary, produced advantages of an utterly new order, so that he soon dominated the earth, and the struggle resolved itself, as far as he was concerned, into an effort to protect himself against the vicissitudes of nature and into an intense and continuing contest between man and man.

Conscious evolution could proceed at an enormous speed. Without waiting for the slow transformation of his surroundings, man proceeded consciously to mold nature to suit himself. He took tools in his hands and thus escaped at once the limits of his mere physical prowess, evading the need for millenniums to alter his body at the slow pace of physical evolution. Finally man began, almost imperceptibly at first, consciously to mold himself.

The later stages are soon told. His great advances came when he controlled fire, when he fashioned ingenious machines, when he learned to record and transmit his thoughts, and when he devised organization. Especially striking was the origin of the art of command and domination by the power of the word rather than of the blow. It was a particularly potent step when he realized that tools could be made that would create more tools. The supplementing of his muscles by controlled energy, from the domesticated animal to the jet plane, gave him extended control over his environment and new ways of attacking his fellows, especially when he found that energy could be released suddenly in an explosion. Finally, his greatest step came when he found that he could understand nature by logic and experiment and that out of his science could come applications that went far beyond chance improvisations, for out of this realization came modern communications and transportation, new and effective ways of altering the surface of the earth and of securing what lies beneath, energy in a hundred convenient forms to do his labor, the partial conquest of disease, and finally the atomic bomb. Moreover, his science extended to himself, and he learned the power that propaganda has to sway multitudes.

So finally we have man, endowed with enormous power over his surroundings and his fellows, governed by his old instincts, predilections, and prejudices, thoroughly at war with himself by new and terrible methods, squandering his precious resources, conscious of himself and of his small position in the cosmos, crawling on a rock that plunges wildly through space, and wondering what comes next.

The wave of barbarism that threatens to sweep the earth is controlled by an organized few who would conquer the world, who hold simple kindly people in thrall and mold them for the conquest, and whose idea of the future is even starker than the past that has been summarized above. Those among them who still hold to ideals of altruism or humanitarianism, who believe they have a new economic philosophy of value to teach the world, are carried along in the mad crush by those who dominate and indoctrinate the entire system. The central group has no doubt as to where we go next; we will continue the rule of tooth and claw, with all its treachery and fury, for that is our nature. Those who accept and teach this terrible materialism, and who also intend to make something out of it, are no new group in the world. But it has now attained new and frightful power with the advance of science and its applications. This is not only because application of the physical sciences has brought new weapons and the ability to move far and fast, to speak to multitudes, and to manage complex affairs by reason of gadgeting aids. It is also and more forcibly because distorted reasoning from misinterpretation of physical and biological facts has led to perilous and perverted social theorizing, and because the sciences that deal with man, imperfectly grasped perhaps in their real attainments, have led to very practical applications in propaganda, indoctrination, deceit, and dominance. The members of this group, in whatever country they may be, not only are completely materialistic in their outlook, but also make a practical working formula out of their philosophical approach to life. It is a new religion made up of the complete lack of religion of any sort, if by the one we mean a way of life and by the other a departure in philosophy from strict reasoning based on physical experience. Its attacks are met today by a wavering defense, for one reason because, in a world pervaded by science and its applications, where some of the old fears are clearly dissipated and others are assumed to be and are replaced only by the fear of men, the old religions have lost some of their power to sway the minds of men.

The group is very practical. It takes the long past path as delineated by man's inquiring mind and postulates its continuance as inevitable in its stark repulsiveness. The law of fang and beak is the only law, in this reasoning. There are no moral codes, no virtue and no honor, nothing admirable in man except his will to survive and dominate by force or cunning. The arts

are merely sublimated survivals of race experiences and superstitions, to be used like everything else on the multitude, for
thus it can be swayed. The gentler virtues are merely holdovers
of what was useful in the ancient struggle for survival, when
mammals found it desirable to prolong youth, and wolf packs or
clans needed taboos to regulate their assaults effectively—holdovers still useful for the same purposes.

The human race will go on, according to this group, fighting
for existence and a place in the sun. New weapons will make the
battle swifter by lending themselves to mass destruction, but
the lack of them need not be a serious deterrent, for there are
subtler ways of dominating and enslaving. Mankind has awakened from its dreams and delusions and now faces the hard facts
of the cold-blooded struggle for existence on a withering planet,
where the survivors will be those who most clearly grasp the
implications of the struggle in its modern dress and their slaves.
Nothing lies ahead but a cycle of conquest and overturn, the
struggle of races followed by a single world domination in
which a war of factions inside the palace becomes the new form,
and may even be learned now. Eventually fuel and minerals will
be exhausted, and man will become a beast again, as the earth
rolls along to its cold end. But in the meantime there are the
zest of animal pleasure, and the satisfaction of being more cunning for a time, and of having the privilege of scoffing from an
eminence.

In practical terms, the prospect is elaborated more than this.
There are corollaries, and it is sometimes difficult to see just
where the pattern leaves conviction and turns to convenience.
The story goes on: Freedom is a myth. Totalitarianism is inevitable. True, it must be brought about by delusion and the
elaboration of seductive goals near at hand, but the only fully
strong system is one with centralized power in the hands of a
small, close-knit group that fully understands and makes sure
the multitude does not, that develops its own rigid code for survival and very effective ways of enforcing it, so that it is the only
group capable of controlling the vast ramifications of man and

nature. Such a control will finally force uniformity of pattern, and even a sort of contentment, where independence of thought will be defined as insanity. There will be two patterns of thought only: one for the people and one for the elect, and both standardized. Truth will be redefined as conformity. There will be no artificial danger of breaking the myth this time by turning to nature for new and unexpected evidence, for the core of the accepted pattern will be that body of knowledge produced by science itself, and no other will be considered the product of a sane mind.

As this group sees it, mankind is today divided into two classes: first, those who postulate an inevitable consummation and jockey for position inside this pattern of inexorable destiny, and second, the fools who do not see and who can be managed, or conquered and enslaved. The most useful of the fools are those who give intellectual assent to at least part of the materialistic story, and who are hence exceedingly useful to preach the doctrine, but who do not understand it all or have not the requisite ruthless determination, and who are still swayed subconsciously by the folklore of their youth or some concepts of fair play acquired among the dreamers, and who are therefore not to be seriously considered as contenders inside the really hard-boiled core. The concept of human dignity is one with the old fanciful codes of ethics and the fading superstitious religions, and these are all passing fancies for the weak and irresolute. The enthusiasm of youth for a fair world, where strong men share their dangers and blessings and work for a common end, is a naïve motivation which can be exploited to bring them under control. All this talk of freedom and democracy is also a passing phase, with plenty of paraphernalia that can be utilized in the pattern of transitory indoctrination, but on its way out. The only enduring philosophy is one that rests squarely on the findings of cold science. On this, world domination can be built by those who grasp its hard sense, and this time there will be no fooling because of the advent of new ways of thought. This is the pattern of a new fatalism—a fatalism whose core is the selfish materialism of conquerors from Alexander onward but whose menace has been intensified by the rise of science and its exploitation by those who understand a little of it only too well. It is the governing philosophy of powerful men, and it is their threat that the rest of mankind faces. The threat is plain, and takes us back to the story that we have come to a fork in the road. The threat is plain; it is that this group will drive mankind down the path to barbarism and despair.

No longer are the forces of disruption engaged merely in underground subversive efforts. No longer are law and order perched on a volcano, the rumblings of which are distant and the eruption improbable and remote. The technique of disruption is the principal tool of a tightly organized band of a million men, who hold hundreds of millions in their sway, and who make no bones about their intentions. They propose to keep the world in confusion until it cracks and then to take over the pieces. There is nothing subtle about it; the air is filled with their raucous vituperation; they enter all the councils but only to confuse; they stop at nothing—nothing, that is, except open war, now.

In their zeal, perhaps in their despair, for every individual is desperate in a state based on intrigue and deceit, even if the state is not, they count on their daring to push closer to the abyss of war than those who oppose their will, and by this daring to prevail and extend their rule.

Their jostling about on the edge of the precipice would have been terrifying enough in any case, in a world shaken by destruction and groping for the path upward amid its ruins. When a new war would involve atomic bombs, and might involve new poisons or the spreading of disease among herds or men, it acquires a new terror. It is not an immediately overwhelming terror, only because there is an equal terror in the slavery that conquest would bring.

Thus there are those among us who cry out that the case is hopeless, and who thus furnish fuel for the fires of those who would prevail by fear. A third war is inevitable, they cry; atom bombs in thousands will destroy all cities and all man's works;

man-made plague will wipe out the race, except for straggling survivors thrust back into barbarism.

There is a fascination in fear. There is a vortex that surrounds the concept of doom. When there is stark terror about, men magnify it and rush toward it. Those who have lived under the shelter of a wishful idealism are most prone to rush into utter pessimism when the shelter fails. No terror is greater than the unknown, except the terror of the half-seen.

This is the threat we must meet. It can be met by preserving and enhancing our strength. But to this must be joined a conviction, a faith, a philosophy that sees more to life than an inexorable fate grinding out a sordid destiny.

Fatalism is no new thing in the world. The threat of the surge of vast hordes actuated only by the lust of conquest is no new thing. But science has entered in two ways, and the threat takes new forms. Science has altered war. But it has also molded the thinking of men. Fatalism is no longer merely a negation; it becomes now an active philosophy based on an interpretation of science as utter materialism. A struggle by the oppressed toward more freedom, a movement based on a Marxist philosophy of revolution by the proletariat, has developed into a police state, dominated by a tight oligarchy and bent on conquest. It is no longer a system based on new theories in regard to the relations of classes within an economy; those who rule now distort their early teachings to meet their current convenience. It is now a tightly organized nation in which the early slogans in regard to the rights of workers have become utterly hollow. What began as a humanitarian movement, ruthless and limited, it is true, but nominally dedicated to the improvement of the lot of the oppressed, is humanitarian no longer. It is merely a state dominated by a few and dedicated to dominating all of us. Humanitarianism has given way to fatalism. All faith, even the faith in the ability of the common man to govern himself, has been discarded. What remains is the ambition of dictatorship, nothing more.

Its central characteristic is its materialism. It takes the teachings of science, as it sees them, and applies them to its program. In the name of that science it denies all decency between men. There is no such thing as international morality; there is no such thing as morality at all, unless conformation by fear to rigid orders can be called moral. This is the new fatalism, embellished by science, and it is the threat we face. The advent of this scientific philosophy has made the threat of fatalistic hordes concrete and coherent. The advent of the applications of science has placed new tools in their hands.

Yet the whole affair is a ghastly fallacy. Science has been misread. Science does not exclude faith. And faith alone can meet the threat that now hangs over us.

Science does not teach a harsh materialism. It does not teach anything at all beyond its boundaries, and those boundaries are severely limited by science itself.

Science merely observes what is apparent to man's senses, describes what it sees, condenses this description into generalized form for convenient use, and thus predicts how material things will function in their interrelationships.

Science builds great telescopes to extend the power of man's vision, and with these it peers into the vast reaches of space and studies how the cosmos is assembled and how it moves. It develops a process of symbolical reasoning, which it calls mathematics, for combining its observations into logical systems of cause and effect. It reasons thus how the stars may have been formed at some remote time by the condensation of cosmic materials, how they may be rushing asunder at prodigious speeds, and whether the space in which they move is curved so that a path long extended would return upon itself. It reasons how the stars maintain their brilliance, and how for this purpose they derive the energy locked in the atom. It predicts how they will cool, and how the vast energy of the heavens will be redistributed. But it does not examine how the cosmos first appeared to be reasoned about. Still more strongly, it is silent as to whether there was a great purpose in the creation of the cosmos beyond the grasp of the feeble mind of man. These things are forever beyond its ken.

Science builds microscopes to delve into the inner recesses of matter, ever more intricate and powerful, calling upon the behavior of the X-ray and the electron to extend man's sight to the very small. It goes beyond this, beyond the minuteness that can thus be revealed, and by interpretation of the meaning of needles on sensitive instruments it explores the atom itself, with its surrounding electrons and its mysterious core, and describes how atoms react upon one another to produce the whole array of chemical behavior and transformations, the physical properties of matter. Thus, it creates new combinations and new properties for man's well-being. It even probes into the internal secrets of the nucleus of the atom and the forces that bind it, and learns how to derive energy from these far exceeding in potential extent the energy from coal or oil. It writes compact and bizarre formulas to summarize what it finds, to interrelate the electrical and gravitational forces that it thus distantly observes, and the inner forces that it by no means yet understands. It speculates as to whether all is cause and effect, or whether there is an element of probability and chance, even in the interrelation of physical things. But when it comes to the reason why these forces exist, what their ultimate nature is, how they came to appear, it pauses. These things are beyond its ken.

Science looks at life. It measures and compares, and it brings to bear sharp instruments to probe the regions of the cell, and microchemical means for unraveling its processes of growth and change. It traces the way by which life evolved on the earth, how new species rose and how they continue to arise. It explores the mechanism by which the simple organism develops, how its pattern was predetermined by its inherited genetical constitution, and how this controlled its growth from the egg to the adult, by mysterious chemical messengers and agents, how it produces enzymes that it may convert materials about it into its own structure and for its needed energy, how it produces hormones to regulate its refined internal processes, how it protects its mechanism against the ravages of invaders and parasites. From its knowledge of the mechanism of life science prescribes means by which man may alter the course of nature about him for his own ends, how he may make two blades of grass grow where only one grew before, how he may protect himself against the onslaught of disease and lengthen his life and his period of effective functioning. It traces the evolution from a primordial cell under the sun to a system of organic life culminating in man, and it teaches man how best to cope with his environment. But it does not speculate as to how the materials and processes that were involved came ultimately to be present, or whether these were chance or were expressly designed to produce a man. These things are beyond its ken.

Science probes into the mind of man itself. It traces how nervous systems arose to connect the stimuli of organisms with their optimum response. It studies how these operate, by electrical impulses that they transmit, and by chemical phenomena along their length and at their termini. It examines how these crude beginnings evolved into a brain, where experience could be stored in the pattern of interconnected cells, and conditioned response could appear to introduce judgment into the responses of an organism to its environment. It studies the functioning of the brain of man, the intricate manner in which it becomes apprised of the conditions of its environment, the relations of the active portions of the brain to the subconscious regions that combine and sometimes contest. It unravels abnormalities, and tells why men react as they do to the conditions they experience. Out of all this it is beginning to teach what it means by a healthy mind, and how to preserve it among the vicissitudes and trials of severe existence. But it does not define consciousness or tell us why there is a being on the earth who can reason as to why he is there. It does not speak with authority as to whether there is such a thing as free will, a choice of actions over and above that dictated by the operations of the mechanism. It does not deal with faith. These things are beyond its ken.

So those who contend that mankind is engaged merely in a
futile dance, a meaningless fluttering over the cruel surface of
the earth before an inexorable curtain descends, with no more to
life than a struggle for a seamy existence, do not do so on the
teachings of science. They do so because they conclude that
the limited observation of our weak senses and their petty aids
encompasses all there is of reality. From such a fallacy come
materialism and the new fatalism now built into a political system geared for conquest.

The threat is now definite. The agelong contest between those
who would build and those who would dominate now crystallizes
into a new form. On the one side are those who see in life only
a harsh struggle, whose fatalism now rests on the materialistic
fallacy that science teaches all there is to know or feel. On the
other side are those who have faith that life has meaning, who
would follow science where it applies, but reach beyond in
aspiration.

Those who oppose the threat are men of good will. They make
up the great company which believes in human freedom. It is
a varied company, for it includes those who adhere to the great
formal religions, those who do not thus adhere but who have
faith, and those who order their lives apart from deep religious
thinking but who have hope for man and believe he is capable
of building a better world.

The great company includes all those who have religious faith
and who delegate all or part of their thinking on the profound
problems of life and morals to a clergy or priesthood specially
prepared for the contemplation of these problems, and who pursue their ways under their guidance, in the organized great religions. It includes those who thus guide, when they do so by a
firm conviction and an accepted theology and code. The great
religions are still a powerful force in the world. The release of
thinking owing to man's new ways of preserving and communi-
cating his thoughts, and the freedom that has come as he has
been spared some of the rigors of nature, have brought a change
and caused a shift. Many have left the great religions by leaving

all religious conviction, but more have left because their philosophy of life has become such that their conviction no longer enables them to delegate to others their personal struggle with the problem of what life is all about. But formal religion is still of vast influence and power over man.

For our purposes, however, we do not examine the thoughts of this great group, diverse as they may be in detail, as to the path that humanity will now take in its great adventure. They see the path, each in his own way, and see it clearly in their faith, and no one who does not belong to their particular company can expound it for one of them, nor should he attempt to do so.

They are all allies, each in his own way, in opposition to materialism. Much as they may differ in theology, a common trait is drawing them together in a solid phalanx against the tide of politically applied materialism that would submerge and destroy them all.

The company we consider includes also all those who ponder about the destiny of the race and order their lives by their convictions, without subscribing, on the one hand, to any formal religion or, on the other, to the thesis of the practical applied political materialists. So it can include very religious men on the one extreme and men of no formal religion at the other extreme, provided they have faith in the dignity of man. In particular it can include essentially simple men who formulate no profound philosophy of life, whose opposition to the ways of the Politburo is merely the reaction of decency to its opposite, whose faith is simple, and expressed simply in faithfulness to friends and in trust that there is meaning to this that they do not attempt to fathom. This whole company has, and can have, no centrally accepted and elaborate philosophy and program, but it does hold firm to certain convictions, even though it does not assemble them into a rigid pattern of thought and action and even though its extreme members subscribe to only a part of them.

The entire company of men of good will believes in the dignity of man, and it shows this in its actions. While it is perfectly willing to admit that man has made a sorry spectacle of himself, and that his ways can be disgusting as no beast ever disgusts, yet it is not willing to describe all of man's actions solely in terms of selfishness, or weakness, or depravity. It holds that out of the course of conscious evolution has come more than the mere elaboration of taboos and codes produced by the survival of the fittest to survive, in the narrow competitive sense. There are such things as noble thoughts and actions. Men can be liked and admired for other reasons than that they are strong or clever. Many an ardent evolutionist may explain all this as the mere diversification constructed by a complex genetic system. Yet even most of these will make friends and on a deeper basis than that of advantage or amusement. The fact that they do so demonstrates that even they see something in their fellow men, rising above the sordid and selfish, to which they would adhere.

The company believes in freedom; it believes passionately in freedom of the individual. Instinctively if you will, and quite apart from its logical analysis of how man came to his present state, it believes that there is something in the free creativeness of individuals that is worth preserving in the world. There is more to the emotions than the mere echo of racial terrors and satisfactions. The evening by the woods fire holds more than rest after the joys of the chase, or release from the trivialities of the noisy city. A symphony can be sublime, and the word means more than mere recollection, or the re-excitement of inherited patterns of interconnection in nervous tissue. A poem can touch truths that go beyond those that are examinable by test tube or the indications of needles on instruments. Moreover, these things lose their elusive grandeur when they are dictated or regimented in any way; and it is hence well, for some undefined reason, that man should be free to create and to think his own untrammeled thoughts. Moreover, the satisfactions and joys of life are more than simply the release of pressure on a nerve, or the excitation of a response that racial experience has classified among the pleasurable because of its survival value.

Not the least of these is the conviction that something new appears, something admirable in its way, because man wills it so, not just because of his constraints but rather because in his freedom he can sometimes rise above them.

The company, or most of the company, believes that there is such a thing as truth—not the mere accepted convention, not the formulation that happens to coincide with a set of arbitrarily accepted and undemonstrable postulates that have become by chance the mores of the species, not even the set of statements that can be supported by experience in the indications of instruments with a probability surpassing that of chance and error. Rather, it believes in absolute truth, undefined, quarreled about among the philosophers, only vaguely grasped, yet sensed and clung to through all the morass of logical thinking.

Implicit, of course, in these beliefs is that there is a possible destiny for man higher than the mere struggle for existence, that a great adventure commenced with the beginning of conscious evolution, that something described as free will influences it, and that it is worth while to carry the experiment through to a conclusion.

Now, the approach of the company to all these things is varied, for the spread is wide. Those who join with their knowledge of the world about them, derived from their own sensations and experience plus those of others of which they admit the validity, a conviction that there is more to life than the summation of all that this knowledge alone teaches regard the presence of higher things than those that result from the straight mechanistic avenue of approach as merely a part of the great eternal mystery, which they do not try to fathom but which they most certainly do not deny. And there are not many thinking men who do not find, in one way or another, a point at which to depart from the immediately appreciable. Certainly there are fewer scientists than there are others who would go so far as to assert that the sum of what is measurable is all there is. Every scientist who gets out of a specialized cubicle is soon led to the boundaries where the inability of the mind of man ever to measure, analyze, or grasp is fully evident. There are plenty who, in their impatience with the loose and easy thinking about them, will assert quite the contrary with vigor and disdain, but they are likely to be the very ones who will give themselves away by a kindly, altruistic, unseen act that belies all that they assert.

It is a diverse but powerful company, determined that there shall be no conquest and dominance by mere force, determined that there shall be government responsive to the will of the governed, determined that ultimately there shall be peace under law, that there shall be an environment in which man's mind can continue to grow. It is the entire company of men of good will.

These great contrasting philosophies of life are now embedded in two systems that face each other across a gulf. The split is not purely geographic, for there is infiltration, on the one hand, and there are multitudes of men of good will behind the iron curtain, on the other. It is not entirely the totalitarian arrayed against the democratic, for there are peaceful dictatorships outside the curtain. But the gulf is there, and democracy is the great strength and asset of those who would work toward a world of freedom. The two systems build weapons, and we have examined the weapons they build, and the ways in which science becomes thus applied, and have looked forward to the nature that war would take on should these systems come to the direct clash of arms. But the subject before us, of the relations of science and democracy and war, is far broader than this.

It is well that we should have perspective on this split, and not arbitrarily jump to the conclusion that all idealism is on our side of the iron curtain. When Russia first threw off the yoke of the tsars there was a great surge of idealism, and it is not all dead as yet by any means. The youth of Russia suddenly saw a great vista opened before them, a world of neighborly helpfulness freed from the selfish struggle for the world's goods, where a man could rise and serve his fellow men by reason of his inherent merit, not through influence, or tradition, or intrigue. They saw a community of interest to replace the grasping narrowness of a capitalist struggle for profits, and they devoted their lives to its development. It is one of the great tragedies of history that these beginnings, at their best as admirable in their aspirations as those of our founding fathers, were seized upon to build a police state, a dictatorship of a Politburo, where all freedom and all kindness are trampled upon. The idealism still lives, no doubt. Even in those who have been drawn into the manipulation of the machine, those who form a part of the complex pyramid of control, there still lingers often a fragment of the old vision, rationalized in strange ways under the discipline of the party. In the great mass of the people, who can tell?

Yet there is no doubt of the menace that we face. The controlling spirit of the ruling class is materialistic and ruthless. It must be faced, and it must be resisted, until it disintegrates because of its own failures and gives place to a system with which we can truly collaborate. When that time comes the Russian people may be different, but they will not have lost their capacity for faith, and they may be ready for democracy. In order that this may occur we need to build where they dreamed, not in their exact pattern, no doubt, preserving individual intiative and avoiding a drab leveling, but preserving the essential core of human decency to which the best of them aspired. To do so we need to build our own system of democracy to its ultimate possibilities and strength.

After reviewing the very explicit ways in which science is applied to ships or bombs or aircraft, we have examined very broadly the motives that actuate the two parts into which the world is becoming completely split, and how the growth of science itself has thrown into relief this clash of philosophies and ideals. We now turn back. It is necessary to examine two forms of government and how science and its applications flourish under them. For the course of the world is to be determined by the relative stability and strength of two groups, and no small part of modern strength depends upon the wisdom with which science is furthered and utilized. This depends in turn upon how well governments function. We shall see that the ideals themselves deeply affect this functioning. We shall find, in fact, that the faith that lies behind the actions of men of good will, the faith on which democracy is founded, the belief in freedom and the dignity of man, are powerful, as faith has always been powerful. We shall find that these form a basis for creating strength in a world that science has altered, a strength far beyond what can be created and maintained by any regimented dictatorship, a strength that can build a better world.

CHAPTER XIV

TOTALITARIANISM AND DICTATORSHIP

"There is no iron curtain that the aggregate sentiments of mankind cannot penetrate." —JAMES F. BYRNES
Speaking Frankly. 1947

Is RIGID TOTALITARIANISM to prevail, gradually drawing all free peoples into its orbit to form an enormous police state ruling the earth? Is democracy to spread, converting others to its tenets, until all men live in freedom? Or is the struggle to continue for a generation or more, perhaps until man learns to live in peace even in the presence of many systems and many philosophies, with the mass opinion of mankind able to express itself and enforce something approaching decency in relations among states, races, and peoples?

More immediate, is the Soviet government to be stopped at its present boundaries, or turned back from its recent conquests? Are the free peoples of Europe to rise and regain their strength? Can a union of nations rest on great power and diversity and thus maintain the peace? Can something resembling the union of vigilantes maintain order in a lawless world until central power arises from the will of all the people to create just law and enforce it? Can we avoid war in the process, or if we cannot, can we survive it? Can the standards of living over the world rise, in spite of the burdens of armament? Will chaos or order come out of the present confusion? Will democracy or totalitarianism determine the trends in the near future? These are the practical, close-up questions in which we confront the broad clash of philosophies that has just been discussed.

The answer depends upon whether totalitarianism or democ-

racy is the stronger in facing the complexities of modern existence, in truthfully interpreting the teachings of science, in ably
utilizing the applications of science, in managing peace or conducting war, whether that war be overt or subversive, hot or
cold. We have already fought with totalitarianism—the totalitarianism of the right—and have found part of the answer. Now
we confront the totalitarianism of the left, and we had best
examine it searchingly.

To make that examination is not easy, for we cannot know
with any certainty what the Communist leaders mean by Communism. Preachment and practice, as far as we can determine,
are wide apart. Thus it is argued often by the heads of the Communist hierarchy that the most modern form of democracy is
embodied in the Soviet Union, and they point to the wording of
the Soviet Constitution as proof. In contrast, we learn from
credible individuals of slave camps, secret police, the sudden
arrest in the night, and all the other paraphernalia of dictatorship and government by force. In these things we see evidence
flatly contradicting professions of democracy as we define it.
The hallmark of democracy, in our definition, is that it creates
government by delegation of power from below; it is the an-
tithesis of single-party rule from above. It is possible that the
time may come when the Soviet system of government will
rely less on force and more on persuasion. But as far as we can
see, that time is not yet.

Communism—the totalitarianism of the left—starts, as we see
it, with the thesis that the workers are oppressed, that they
have no hope of redress through the ordinary processes of government, and that they should therefore revolt. Moreover, having seized all property and all power, they should forever prevent
the accumulation of property or capital by any group, for otherwise they would again be reduced to bondage. So there must be
a dictatorship of the proletariat, or a retention of all power by
the working class, with all property owned by the state, the state
controlled by the workers, and the state in turn controlling all
things.

The revolution in Russia more than thirty years ago, as a result of which totalitarianism of the left menaces the world today, was an explosion relieving pressures built up by abuses and grievances over many years. That long history of oppression and suspicion, of grudging reforms followed by reversion to reaction, is beyond our concern here. The revolution in which it culminated might have led in time to a democracy of our sort, but it did not. A brief attempt at democracy succumbed to an opportunistic seizure of power by a tightly organized minority that was ready and ruthless, and no true democracy emerged. The elements of understanding that might have made true democracy possible were not present in the common people. One cannot take the top off a volcano and expect a quiet spreading of a salutary equilibrium.

Many factors were involved in the chaos that followed 1917 —factors ranging from illiteracy and famine inside Russia to suspicion and armed intervention from outside, from abortive efforts at reform and stable government of more or less conventional kind made by enlightened, though inexpert, leaders to the ultimate consolidation of power by skilled revolutionaries. Hence the revolution led ultimately not to the placing of power in the hands of all workers under a general franchise, as had been the profession, but rather to the retention of control by the revolutionary group. There was nothing necessarily disastrous in this move; in fact, it might well have been the only move that could have produced and nurtured a growing system of democracy in an unprepared, heterogeneous people. The general idea was that of a central party of a few millions, to which would be admitted the most competent youth of the land by competitive trial that would demonstrate their capacity to learn and to function in a political organization. Within the party all the machinery of democracy would function, with free elections, open criticism, laws made by legislatures of the party members, and an executive branch chosen by a pyramidal system of selection within the ranks. There was no inherent denial of the democratic form—at first—for a limitation of the franchise to those who are qualified under fair tests is a part of every democratic system, and sharp limitation appeared necessary and reasonable—at first.

But even the party was not ready. On the part of those who jockeyed for political power there was no real reliance on the principles of democracy, and no intention to allow democracy to operate. The idea of free elections was lost because of the concentration of power in a clique, which then ordered the process for its own perpetuation. This led inexorably to totalitarianism and dictatorship.

This inexorable development lies at the root of the difference between preachment and practice that confuses efforts to examine the position of Communism in the international situation today. Verbally, the Communist thesis professes to favor the principles of democracy—of democracy in the sense in which even democratic nations understand the term—even though the governmental mechanisms it sponsors are a far cry from the representative processes we associate with democracy. But in action, at the present time and for years in the past, the dictatorship that has grown out of Communism in Russia refutes this profession. Whatever the verbal profession may be, the operations of the Communist dictatorship as we experience them appear flatly to deny the principles of democracy. They indicate an utter cynicism as far as our system of government is concerned—a belief that our whole scheme, of elections and what not, is a sham, that the only real power resides in those who have money, who can buy the press or legislatures, who can determine the outcome of elections and thus rule. This country by definition is a capitalist country ruled by its wealthy class. Moreover, and here is the rub, they argue that we fear a revolution ourselves, and that hence we are convinced we must by all means stamp out the source of revolution, the Communist state, before it engulfs us. Under this theory, no compromise is possible—either the workers' revolution that started in Russia will engulf the world or it will be overcome and the whole world will return to the domination of a self-perpetuating privileged class that will hold the workers in subjection. It sounds fantastic, but that is the thesis.

The Communist state that centers in Russia is not merely Russia or the USSR. It is the entire body of Communists everywhere who are linked into its system and governed by the dictatorship of Moscow. To them, the state is uppermost, and the survival of the state is all. To it must bow all other desires and aspirations, to it even religion itself must yield, to it must contribute all art, all literature, all thought. There is no relationship between men that is valid save that of rigorous discipline in service to the state, which does not hesitate to purge or destroy, which balks at nothing if only the state be preserved. There will be no rest until that state rules the earth. Its deadly enemy is the association of great democracies, for these are the modern means through which capitalism aims again to enslave the worker. The democratic state, therefore, like any other competition, is to be overthrown when the opportunity offers, preferably by revolution from within. This Communist state permeates every country, our own included.

We know, of course, that we have Communists in our midst. There are several types. First there are out-and-out emissaries of Moscow, their converts and their dupes and those in their pay, placed here to embarrass us, as they are placed all over the world for the same purpose, under the strict discipline of Moscow, and fearful of departing from the party line in the slightest iota, for terrible would be the retribution. Then there are those who feel that the outcome is inevitable, that democracy will fail, that the capitalist system is doomed, and who therefore wish to be on the side that will soon triumph. They have no objection to the country's being ruled by a clique; as they reason, it is merely the wrong clique that rules at the moment, but it will soon change and they will join the new one early. We may not have many of these, not nearly so many as in countries where the issue is more in doubt, but we have some of them.

These two groups above do not account for all our Communists. There are those who are Communists from an emotional standpoint. They are oppressed by the world, or they think they are, and they are frustrated in their aspirations, or they believe that artificial barriers are all too prevalent. To them all democracy is a mockery—the power of money really controls and all else is sham. There is no recourse except to overturn the existing system, and the ills of humanity cry out for a change. They do not reason very much as to where they themselves would lead, and seize upon Communism merely as the obvious party of protest.

Even yet we have not included all of our Communists, few though they are in percentage. The greatest number are not Communists at all; they merely think they are. They are extreme humanitarians, or they delight in that appellation. They would do something for downtrodden man, and they have a sequence in mind, running all the way from reaction to liberalism, which proceeds from the monarchy, through democracy, and the so-cialist state, to the most liberal of all, the communist state. It is fuzzy thinking, of course; there is nothing further from true liberalism than modern Communism as it is practiced, but the words delude them just the same. There are fewer of these queer souls than there used to be—the spectacle of the Russian dicta-torship in action has cured many. We are left with the incurable.

There are also those who pose, who would embrace almost anything rather than be thought to be a Babbitt, those to whom it is a cardinal sin to be ordinary. They like to be thought differ-ent, or daring, or what not. This pose is not so fashionable as it once was, among our lunatic fringe, but we have every variety of crank, and their lack of numbers is offset by their conspicu-ousness whenever they can attract attention.

There is one more group, and it is important, not because of its numbers but because of its quandary. When the revolution occurred in Russia not all of the idealism that was its accom-paniment was located there. Throughout the world there were those who grasped at the beginning of a new era, the advent of a system that would succeed capitalism with all its faults and construct a better society. They were the liberals of the extreme left, who would jump at once to an ultimate solution of all our ills, a solution with which we might by no means agree, but on a basis that we could respect. They were the extreme idealists, who held to an advanced philosophy of very simple relationships among men and expected them to become embodied in and exemplified by the new movement and the new nation. Now they are in a quandary. The nation has become a dictatorship that grinds to atoms the human rights they cherish. In England a socialist government has taken hold and moderated in the process, and they cannot adhere to it because it does not go far enough. Yet they hold to their old tenets and do not know where to turn. Some of them rationalize the whole affair; the Russian movement is still a symbol to them, and they adhere to it even when doing so is absurd and support it even when support involves playing the game of the Russian party against their own homeland. These we can classify, for they are part of a conspiracy. But there are others, still loyal, still extreme liberals, baffled by the turn of events. As we contend with a rival system, as we call a traitor a traitor, as we weed out those who would be aides to an attempt to overthrow our democracy by force, we need to leave a place for the honest liberal of every sort, provided he would work out his destiny by orderly processes. We should not qualify our acceptance of liberalism by its degree, as long as it is loyal. Yet, even as we do so, we cannot compromise with the dictatorship that takes the name of modern Communism, which is our adversary in the present struggle.

In even the recent past, it has been dangerously easy to be led astray and to discount the power of the Communist state. That state has very real power, very real strength. That strength is really of two kinds—the beneficent and its opposite. That the regime in Russia has brought hope and opportunity to many no one will deny. For generations under the tsars, the children of peasants knew if not outright serfdom, at least the complete lack of a chance to rise by their own efforts. Social mobility—which we of the democracies place at the highest premium—was an impossibility under the system that 1917 overthrew. Since

1917, social mobility, in terms of at least a chance to advance oneself through hard work, has been made possible, even if at the price of conformity. And even at that price, it breeds gratefulness and support of the system—and these mean strength.

What of the other strength the Communist state possesses? This lies in discipline—it is the other side of the medal of conformity. The Communist state is not the dictatorship of a single man—even though at any instant all real power resides in an individual, for it is so arranged that if that individual loses his grip, he will be replaced by intrigue or by murder, and a new dictator will arise to control the system and proceed with the plan. The strength of the Communist dictatorship lies also in its discipline, for it can control all of its agents by harsh punishment and the fear of it and can send them among its enemies and into danger and depend upon their following orders, even to the extent of great sacrifice in the cause.

We cannot here trace the ramifications of all this affair: whether the system contains the seeds of its own disruption, whether its satellites can acquire the strength to defy it, whether it will evolve into a new pattern of despotism, or whether, indeed, as we hope, it may ultimately come to rely more on persuasion, less on force, and so become what we consider a more democratic system. This book is concerned with the influence of science upon war, and the interrelations of science and government where the potential pressure of war is a determining factor in our future. We can examine briefly the way in which science will operate within such a dictatorship, and there is no surer index of the weakness of the Communist state than the one afforded by that examination.

The weakness of the Communist state resides in its rigidity, in the fact that it cannot tolerate heresy, and in the fact that it cannot allow its iron curtain to be fully penetrated. All these things, vital to totalitarianism whether right or left, are fatal to true progress in fundamental science. They are not nearly so fatal to the application of science, but they are a severe deterrent to even the healthy growth of this along novel lines.

Dictatorship can tolerate no real independence of thought and expression. Its control depends entirely upon expressed adherence by all to a rigid formula, the party line. Its secret police must be ever alert to purge those who would depart from discipline and think their own thoughts, for departure would soon lead to a vast congeries of independent groups defying central authority, and the system would break.

No true art, no true fundamental science, can flourish long under such a system, no matter what the individual genius may be. Musicians, some of the finest and most creative in the world, are disciplined because a commissar does not like their music, and they bow to the inevitable, apologize and admit their errors, and promise to conform. A great scientist is torn from his post and sent into cold exile because he dares assert that there is validity in the modern theories of genetics, contrary to the state teaching that environment is all-controlling, and he is replaced by a charlatan who will see to it that the state theory is taught to young scientific disciples and that all research is based on a blatant fallacy and an unsupported hypothesis.

Under such a system art will eventually become merely a dull adjunct to monotonous propaganda. Science will eventually become a collection of superstitions and folklore. Men of genius will languish and succumb to discouragement. It will take time. The spirit of a great people that has produced sparkling figures in the past in music, in mathematics, in art and science, will not be broken in an instant or a single generation. The spirit of genius is strong, and it has always risen through obstacles to make its presence known. But in the long run, that very science, on the distorted interpretation of which is built the philosophy of the group that rules in Russia, will become a sham and a mockery.

The situation is not nearly so clear when we come to the application of science, to that interplay between scientists and engineers, industry and government, by which the fruits of new advances in fundamental science are made available for practical use in manufacture or agriculture or war. To throw light on that matter we shall digress to chart illustrations from history in the field of atomic energy and in applied science.

In the years since the war, the honest and determined attempt of this country to place the atomic bomb and all its works under genuine international control has failed. Not quite all of the world was ready for it. It was a sound move; it will stand in history as the wisest step ever advanced by a great nation in its full strength and in the flush of victory. It was well thought out, to avoid the pitfall of false reliance on ineffective controls, and it expressed clear grasp of the practical relationship of international-control activities in a world still steeped in intense nationalism. It remains in abeyance and awaits a better day.

It failed because an essential element of any system of control that is not to be a delusion and a snare is an international system of inspection that works. The rulers of Russia could not accept such a system, for it meant a penetration of the iron curtain in no uncertain manner. Once that curtain was pierced, so that international committees could move freely about within the country, consulting with those who operated factories and laboratories, examining into the flow of raw materials and finished fission products and checking against their diversion, uncontrolled by police as long as they adhered to their proper affairs, the door would soon open wide. The population would learn the true state of affairs in the world. Disaffected citizens would find means to cross the borders, their families with them, and escape the clutches of the secret police. Industrial units would not be completely subservient to central control. Controversy, departure from the party line, would spread, rigid discipline would be lost.

Those who rule Russia would not take the risk—even though it was probably desirable from the Russian standpoint, even in a strict military sense, that the atomic bomb be somehow removed from the armament of nations. There may have been a variety of reasons why the rulers of the Soviet Union would not take the risk. One of the reasons may well have been that the rulers of Russia did not want to follow the suggestion that they personally commit suicide, for one does not lose a high post there and retire—he loses the post and dies. The chain of events set in motion by true international inspection would probably, sooner or later, have challenged the absolute power of the central dominating group. So the atomic-energy proposal and all proposals for controlled armament failed, and we became committed to an armanent race in spite of ourselves.

But—and here is the weakness of rigidity—the iron curtain operates in both ways, to keep people in and to keep ideas out, and among the information and opinion it excludes there will be much that is needed for great progress, not only in fundamental science but also in its applications. Scientific publications of all sorts, pure and applied, will cross the border, but this is not enough. Unless they are accompanied by free analysis and discussion that cuts across international borders they will lose much of their value. As the Soviet Union places complete bars upon effective international scientific interchange, and to the extent that it regiments scientific discussion within its own borders, it places upon itself a substantial handicap in any technical race, a greater handicap than is readily appreciated by us who live in freedom. Here is a practical illustration from applied science.

Some twenty years ago a group of Russian scientists, headed by Professor Joffe, who was then a great figure in physics, believed they had made some new discoveries in fundamental theory and practice on insulating materials. After a year or two of reduction to use in a laboratory at Leningrad manned by a highly competent group of young investigators, these discoveries appeared ready to be put to use. They promised on their face to revolutionize electrical insulating materials and hence to be of considerable commercial importance. The Russian government secured patent applications in this country ,and elsewhere, for valid patents could be obtained here by Russia, and still can, for that matter, even though the reciprocal privilege is meaningless. The discovery and invention were announced in lectures here and elsewhere. Industrial groups in this country and Germany were formed to develop the possibilities, and it looked as though the cost of electrical machinery and power might thus be reduced significantly. Great stacks of laboratory records and tests made in Leningrad were made available, experimental work was started, and then after a time it stopped. There was absolutely nothing in the matter whatever; the theory was fallacious, and the insulation prepared in accordance with it simply did not insulate any better than commonly available materials.

Now the point of this story is that there was no intent to deceive on the part of the Russian scientific group. They were completely honest; moreover, they were able men and women, and personally they were likable and ingenuous, the same sorts of individuals as hard-working, devoted, professional men everywhere. Yet for several years they had followed a false lead, into all sorts of ramifications, without once finding out that the structure they had erected was utterly unsound. This was many years ago; controls have tightened enormously since, and the iron curtain has come down hard. But even at that time this able group was operating in isolation, and its strange adventure was the result of its remoteness from unbiased criticism and of the regimentation of its internal thinking. The extent to which, in the absence of truly critical examination, a competent group can follow a false scent is appalling. Science and its practical application proceed by trial and error, with tentative experiment and hypothesis, with the winnowing of chaff by competition and criticism, and the gradual formation of a sound line of advance by the survival of the fittest. When science enters upon an artificially protected but absurd line it can produce monstrosities, just as physical evolution can produce a toucan with an absurd beak or a dinosaur so heavy that he becomes bogged down in the mud. That is what happened in this case, and the instance is undoubtedly now being repeated in exaggerated forms.

Yet even more than unrestrained criticism is needed for healthy advance in fields that involve many skills and techniques. Interchange and collaboration on a give-and-take basis are also needed. The larger and more complex the effort, the more this becomes essential to progress. We return to the history of atomic energy for an illustration.

The atomic-energy program during the war was to the nth degree the sort of collaborative program that was impossible at that speed outside democracy. It was not merely a matter of new physics and its incidental application—very far from it. True, some of the finest theory and experiment in the physics of the atom was involved, calling for ingenuity and resourcefulness, mathematics of a higher order, and judgment such as can be exercised only by men who are utter masters of their craft, and all this was performed magnificently by the physicists of this country, England, and Canada, in close interchange and at unprecedented speed as the program of application proceeded. But then came the heaviest part of the job. It involved new, dangerous, and complex chemistry, the most refined sort of chemical engineering, industrial organization that tied together effectively the performance of ten thousand firms which supplied parts, built new and unheard-of devices, constructed and operated enormous plants where the whole affair functioned as an interlocked unit. It involved the joint action of diverse groups, theorists, engineers, instrumentalists, designers, in the production of fissionable materials and in the construction of the bomb itself. It involved management that reached a new order of functioning to bring all these elements together in an intense race against time, where nerves were bound to be frayed and patience short. It involved collaboration between military and civilian organizations, with their widely different approaches to organizational rules and systems. It required integration of new elements into the strange structure of government and competition with other programs of highest priority in the maelstrom of war.

The keynote of all this effort was that it was on an essentially democratic basis, in spite of the necessary and at times absurd restrictions of secrecy and the formality that tends to freeze

206 MODERN ARMS AND FREE MEN

any military, or for that matter governmental, operation of great magnitude. If certain physicists thought the organization was functioning badly in certain respects, they could walk in on the civilian who headed that aspect of the effort and tell him so in no uncertain terms. They not only could, they most certainly did; and the point is that there was no rancor, and old friendships were not destroyed in the process. If civilians and military disagreed, as they often did, there were tables about which they could gather and argue it out. Punches did not need to be pulled, and no one kept glancing over his shoulder. If there were international misunderstandings between allies, and there were, they could be frankly discussed, sometimes with more heat than light, but always with a prevailing atmosphere of genuine desire to arrive at the conclusion that made sense and that best got on with the war. If a young scientist had an idea he did not have to pass it through a dozen formal echelons and wait a year; he could talk it over with his fellows and with superiors of accepted eminence in his own field and be sure it would be weighed with unbiased judgment by men of competence. The system worked and it produced results.

Now the Nazis, the totalitarians of the right, were also trying to make a bomb, and they failed miserably. They had the same opportunity that we did; the starting gun for the race went off with the experimental confirmation of the phenomenon of fission in 1939, and this was known all over the world. At the end of the war it was found that they had made little progress; they had not accomplished five per cent of the task that was successfully brought to a conclusion in this country, with the collaboration of England and Canada. That they were far behind in the race was not known until the Alsos mission revealed the true state of affairs after the fall of Stuttgart. Until then we felt that they were close competitors and even that they might be six months ahead of us, which would have been disastrous. We maintained essential secrecy well in this country, and gave the Nazis credit for being equally adroit, so that their possible progress was always a burr under our saddle, until thorough intelligence work revealed the truth, fortunately in time, so that the European campaign could be carried through to its finish in the light of accurate knowledge of an important consideration.

Why were they so far behind? Bombing and the destruction of needed industrial facilities account for some of the lag. Limited availability of critical materials accounts for some. But the real reason is that they were regimented in a totalitarian regime. There was nothing much wrong with their physicists; they still had some able men in this field in spite of their insane rape of their own universities. They were not as able as they thought they were, or as they probably still think, for their particular variety of conceit is incurable. But they were able enough to have made far greater progress than they did. Their industry certainly demonstrated that it could produce under stress such complicated achievements of science and engineering as the jet plane. Their Führer and their military were certainly keen for new weapons, especially a terror weapon with which to smite England. Yet they hardly got off the starting line on the atom bomb.

A perusal of the account of German war organization shows the reason. That organization was an abortion and a caricature. Parallel agencies were given overlapping power, stole one another's materials and men, and jockeyed for position by all the arts of palace intrigue. Nincompoops with chests full of medals, adept at those arts, presided over organizations concerning whose affairs they were morons. Communications between scientists and the military were highly formal, at arm's length, at the highest echelons only, and scientists were banned from all real military knowledge and participation. Undoubtedly the young physicist who penetrated to the august presence of the Herr Doktor Geheimrat Professor said *ja* emphatically and bowed himself out, if he did not actually suck air through his teeth. The whole affair was shot through with suspicion, intrigue, arbitrary power, formalism, as will be all systems that depend for their form and functioning upon the nod of a dictator. It did not get to first base in the attempt to make an atomic bomb.

This all became clear when research teams were sent into Germany after the war to study such problems as the German handling of strategic materials, manpower, propaganda, military organization and co-ordination. All of these teams were amazed. The Germans had made astonishing mistakes in every sphere of action, which continued throughout the war without correction. They had made thousands of little mistakes as well as the big ones.

It had been popular with us for many years to complain about the wastefulness of the democratic system of government. We had come to expect fifty cents' worth of results for every dollar paid in taxes. Benevolent dictatorship was thought to be efficient by comparison, and it was held that a dictator could get results much more generally and cheaply. What we found at the end of the war exploded this myth for all time. Every team we sent returned convinced that the democratic system is clearly more efficient, dollar for dollar and hour for hour, than any totalitarian system. The criticism applies to the German system before the Nazis took over, for it was autocratic long before that, with only a brief interregnum of attempted democracy, which failed because the people were not genuinely seeking freedom; the Nazis merely developed and tightened the system they found in existence.

The teams also found the key to the difference, the key to the effectiveness of the democratic system. It lies in the fact that in a democracy criticism flows both ways, up as well as down, and we shall have more to consider on this point later. Here it may merely be noted that this factor, plus democracy's ability to call to the aid of government at need the views and judgment of experts in any field, who operate, when thus called, with complete frankness, far more than offsets the apparent looseness of democracy and renders the whole structure live and virile, changing and adaptable, rather than frozen into a pattern where any absurdity or any incompetent can persist if politically entrenched.

The type of pyramidal totalitarian regime that the Communists have centered in Moscow is an exceedingly powerful agency for cold war. It is capable of holding great masses of people in subjection, indoctrinating them in its tenets, and marshaling them against the free world. It can force its people to enormous sacrifice and thus build great quantities of materials of war. It can educate large numbers of men and women in science and engineering, construct far-flung institutes, mechanize agriculture, and ultimately create mass production of the manifold things it needs. But it is not adapted for effective performance in pioneering fields, either in basic science or in involved and novel applications. It has many of the faults of the German dictatorship, magnified to the nth degree. Hence it is likely to produce great mistakes and great abortions.

This does not mean that we can flatly disregard the Communist state and cease our advances in the techniques of war. The Communists can copy and improve, and a whole mass of scarcely developed techniques remains from the last war as material for this process. It does mean that we must continue to break new ground, and that we can do so with our heads high, for we have a system essentially adapted for the purpose, if we do not distort it or sacrifice it to false gods of fancied efficiency.

The Russian people are rugged, long-suffering, tough. They are likable. They have among them men of imagination capable of giving much to the world. They are capable of going far even under dictatorship. Should they escape from it by some miracle, should they become in some manner a part of the free world, they could be a great people, with an important share in some ultimate community of nations. In the meantime, as we contest doggedly with the system that controls them, and hold its power in mingled fear and contempt, we should not mistake the superimposed hierarchy for the people themselves. The sound way in which to carry on the struggle is to preserve without question a system that is better, better even for making war, and far better for advancing science and gathering its fruits, and we should let the world, and the people of Russia when we can, know about it in its genuine characteristics and form.

As we do so, as we conduct the disagreeable contest of the cold war at enormous disadvantage, we may be reassured. The system with which we contend cannot possibly advance science with full effectiveness; it cannot even apply science to war in the forms it will take in the future, without mistakes and waste and delay. Moreover, it cannot possibly alter its pattern and become fully effective without at the same time becoming free, and if it becomes free the contest is ended.

CHAPTER XV

DEMOCRACY

"The best answer to Communism is a living, vibrant, fearless democracy—economic, social, and political. All we need to do is to stand up and perform according to our professed ideals. Then those ideals will be safe."   —WENDELL L. WILLKIE
One World. 1943

THE ESSENCE of democracy is that government is responsive to the will of the people, that it is their servant and not their master, that the state exists for the benefit of those who create and support it.

Accordingly, those who govern must periodically receive a nod from the people, must have their assent to continue. If the democracy is to be genuine, this assent must be without deceit or coercion. There must therefore be unregimented discussion, a free press and radio, freedom of assembly, and full protection of minorities and their spokesmen, even when what they say is obnoxious to the majority of the people. There must be in the whole structure freedom of criticism, so that the individual citizen, no matter how humble, may criticize without fear the acts of those who govern, no matter how highly placed. Criticism flowing up as well as down is an essential element. The police power of the state must be under the control of individuals directly responsible to the electorate, for force and intimidation must be absent and minorities must be protected in their rights as long as they obey the law and do not themselves seek to prevail by force.

The strength of democracy lies in the manifold blessings of freedom. Some of them we have discussed, and we shall discuss others. Here lies our enormous strength, in its full development dwarfing the potential strength of any system of regimentation.

The weakness of democracy, on the other hand, lies in its lack of rigidly defined stabilities, in its prevalent confusion, and in the danger that it may be corrupted for selfish ends. Specifically, that weakness lies in the danger that a selfish minority might seize the mechanisms of democracy, deny its fundamental tenets, and coerce the majority, or that even a majority in legal control might forget the principles laid down in the Constitution and Bill of Rights and govern tyrannically.

Various fears of this sort have been uppermost at different times. The greatest fears have been two: first, that rich owners of property would combine and by the power of money subvert legislators and governors, and by thus manipulating the machine reduce those who labor to a new form of serfdom; second, that those who labor and have little property would merge under the control of demagogues, seize property and the means of production, and thus destroy the basis of prosperity and make the nation an easy prey to conquest from without or by a dictator from within. Neither of these fears is by any means imaginary; other democracies have succumbed to one or the other of these causes, and in the course of the short life of this nation one or the other has several times threatened it with disaster. Thus far the nation has resisted distortion and has grown in the strength and genuineness of its democracy. The great question is whether it can continue to do so, in the face of a new and sinister element of disruption on a rampage in the world. And this is still a dual threat: on the one hand, a clique might seize power by intrigue and so might disrupt our processes or destroy our economy; on the other hand, the process of resisting the first threat might force our system into a form where genuine liberty no longer exists.

Today the lines are less sharply drawn, the extremes are less far apart than they were at times in the past. Social legislation to ameliorate some of the harshness of the competition for the country's goods has been generally welcomed; strong organization of labor is an accepted element in our affairs; there is much less conceit among those who forget that great private wealth or power carries with it social responsibilities. Yet the dangers are still present. On the one hand, there is hazard that the public purse will be exhausted and the national credit wrecked by extravagance for inherently good or merely selfish purposes, and, on the other hand, hazard that reaction entering as a preventive or result will cancel out our gains. If the people in conscious control cannot keep the economic ship on an even keel and forging ahead, there is danger that some new control, not truly democratic, will enter or be called in to steer it.

Those who founded the country, in the colonies, in the constitutional convention, in the struggles over form that continued for many years, recognized these threats clearly enough. They feared, above all else, that a general franchise placing the control directly in the hands of all the people by the ballot would soon result in the rise to power of demagogues and fanatics, the abandonment of all discipline, the seizure of property by the mob, and chaos that could be resolved only by the strong man on horseback who would take over the army, restore order, and make himself tyrant.

So they created an ingenious scheme for setting up bulwarks against this catastrophe—separate executive, legislative, and judicial branches; a Senate elected indirectly by the legislatures of the states; a President similarly indirectly elected by an Electoral College and elected for a specified term so that he could not be removed at the whim of the legislature; a bilateral Congress in which the presumably hasty acts of a House directly responsible to the electorate would be checked by an indirectly responsible Senate; a veto power residing in a President who would be, at least for an interval, beyond the reach of popular clamor. More important, they consolidated these provisions in a Constitution; backed it with the Bill of Rights, which states explicitly the ways in which the individual citizen should be free from coercion or abuse by those in power; created a Supreme Court appointed for life, which soon undertook the interpretation of the basic law as its primary duty, to the extent that it could annul an act of Congress if it invaded these constitu-

tional rights; and established an Army and later a Navy sworn
to uphold the Constitution of the United States. Some of these
checks and balances have been whittled down with time, powers
formerly indirect have been more directly reposed in the people,
but the essential core remains.

The rigidity of the system at times becomes embarrassing, as
when the President and Congress become deadlocked, and it
might be exceedingly embarrassing should a President collapse
mentally when in office. Our reliance on a written constitution,
with an independent judiciary sworn to uphold it, provides
another point of rigidity that might prove serious were the
rigidity absolute. Yet we deliberately chose the less elastic system in our early days, and we have adhered to it with remarkable
consistency and found ways of mitigating its potentially static
nature.

Over the years the Constitution has become a more flexible
instrument, modified in interpretation to meet changing conditions of the times, and this is well. The courts take on a severe
burden as they thus extend interpretation of our fundamental
law in a society rendered more complex in its instrumentalities
and relations by the manifold applications of science to daily
affairs. It is no light task to hold justly the balance between man
and man and to protect the citizen against improper coercion
by his government, irrespective of his wealth, position, or beliefs, as new and intricate relations and organizations enter the
scene. It is not easy, without making the structure of law enforcement impotent, to uphold free speech in the presence of a
conspiracy to use our liberal practices to harass or even to overthrow our form of government and to set up a form of government that would deny free speech as well as the other freedoms
in the Bill of Rights. Yet the performance of this whole task with
dignity, maintaining the power of the judiciary without en-
croaching upon the prerogatives of the legislature or the executive, is an essential aspect of our system of government. We have
three great branches of our system, the executive, legislative,
and judicial. Every day we are reminded that the effectiveness

of the first two is essential to our welfare. The responsibilities of the judicial branch are no less arduous and no less determining of our progress. Yet they are more subtle, and by the very nature of their duties the judiciary are shielded from the penetrating light that plays upon the other two. Freed from harassment, remote from direct influence of public opinion, the judiciary nevertheless shape our path in the maze of law. They interpret the Constitution as the scene shifts and protect us against the rigidity that might otherwise bind our sinews, which rigidity, because of their presence and their function in a changing world, is apparent rather than real. The extent to which we escape rigidity without becoming confused depends upon the skill with which they perform their function.

Our situation in these respects differs greatly from that of Great Britain, with its executive branch a direct agency of Parliament and its reliance on the great body of common law and tradition in place of a specifically written constitution. But Great Britain has its monarchy and its House of Lords, now reduced in power, and in quiet times an elaborate formality, but representing not mere tradition but continuity and stability, and furnishing a genuine safeguard against political chaos in the event of unresolvable deadlock. It is not to be forgotten that the oath of the British armed forces is to the King. We have no monarchy, and wish none, and for this we pay the penalty of widely separated branches of government and risk the inconvenience of rigidity, relying upon our judiciary for the key role of maintaining orderly progress in a world of change.

The founders also made no provision to prevent the confusion that ensues when there is a multitude of splinter parties clamoring for power and influence, the sort of thing that has at times rendered France impotent because it could not form a stable government having sufficient public backing to accomplish disagreeable tasks such as imposing adequate taxes—the confusion that rendered France pathetically weak in the last war. But the two-party system arose in this country, and this furnished much of the added stability that was needed as some of the original safeguards weakened. As time has gone on, the two parties have not tended to go to extremes of left and right, with consequent wide swings as power shifted; rather they have both tended to follow the trends of the times and have approached one another in characteristics and basic policies, until they are distinguished by the electorate more on the basis of performance—or, admittedly, even whim—rather than on that of platforms and promises. This is a stabilizing development, and it does not prevent gradual alteration in our system and policies, for when a measure has evident wide public support both parties swing toward it. The way in which the parties themselves operate is hardly in accord with purest democracy. It is in some ways a crazy system, but it seems to work after a fashion.

The first crisis came soon after the system was inaugurated. Many of Hamilton's Federalists were honestly convinced that a fully democratic system would soon be followed by the seizure of power by the mob, which would raid the treasury, destroy all systems of banking and the like, and wreck the whole structure. To prevent this, and to secure power themselves of course, they sought a bonding together of men of property to control affairs by their combined control of the means of producing and interchanging wealth. In the minds of its extremists, this oligarchy would see to it that only those who would subscribe to their thesis were admitted to power, would indoctrinate their sons in its manipulation, and would entrench them by absolute inheritance and the power of money to create more money. They would control legislatures, directly by bribe perhaps, but more by bribes cleverly interpreted to be respectable, and by the lure of position, acceptance, and prestige. We do not need to trace the long history of this undertaking and of its failure. This political philosophy has by no means disappeared even today. It had a new lease of life in the nineties, when the enormous corporate aggregation, the trust, furnished new means for its furtherance. But it is no longer widely held, for the proportion of true liberals among men of large means is as great as elsewhere, and thoughtful men have come clearly to recognize that this philosophy is more likely to destroy than to preserve democracy. It is no longer an imminent threat, even though in the minds of reactionaries it may be merely smoldering and awaiting another opportunity.

Yet Hamilton and his Federalists were right in a way. Democracy, our democracy, was weak in its early days, tragically weak and faced by the grasping states of Europe. Confusion abounded, and there was clear danger that a weak federation would succumb or split unless some strong internal movement centralized power. They seized upon the only apparent stabilizing force and made it work in days of peril. That stabilizing force, in its worst form, is a conspiracy of the money power to control. In its best form it is an alliance of men of outstanding ability to produce equilibrium by informal concerted action, upholding one another's hands for the purpose, joining to resist the demagogue, and placing the safety of the structure above the secondary policies and programs on which they might contend vigorously. There is no doubt that both forms existed in our early days and that the presence of the second curbed the threat of the first.

The other threat, the converse of the first, was starkly demonstrated, a scant few years after our government was founded, in the French Revolution, with the attempt to introduce a government of the people, which produced first chaos, then, under pressure of attack, a dictatorship, and finally an empire. That threat has appeared several times in acute form in our own history, but we do not need to trace it in detail, either. At times a mob, or something closely resembling it, has nearly taken over and started to wreck national or state credit, distribute the land and the tools by one artifice or another, and produce a heterogeneous scramble for the immediate instead of a sound system aimed at future prosperity. This has not succeeded either, in part because of the fact that when a man acquires power he often becomes sober, and there are always cool heads about, even in the wildest clamor. Thus, for nearly two centuries ultimate power in this country has indeed resided in the people and in public opinion, and they have not yet destroyed the system that gives them freedom. Power has been abused, as it always will be abused, and we have been exceedingly foolish and short-sighted many times. But the system has grown in strength, and it is more stable than when it started.

Jefferson himself, for all his advocacy of democracy, felt that the system could endure only so long as the majority of the people were upon the land—farmers with property to preserve. He was wrong; it has endured far beyond that time, until we are today a great industrial nation. But his formula for prevent-ing disintegration was correct; he believed that education, incul-cating in all the population an understanding of their world, could provide a bulwark against disaster. It has done much of that, and we shall have more to say about it presently.

The early experiments with systems of democracy were swayed by the conviction, not yet entirely lost, that democracy is forever and inherently weak, and only that state has strength which is controlled in one way or another by a relatively small group, an oligarchy of position or wealth. The enormous strength of democracy had not yet appeared; it came to full development only with modern communication and transportation and was completely exemplified only in the last war. We have outgrown the need for the tight internal alliance of men of wealth. We will never outgrow the need for an unselfish alliance among men who place the stability of the republic above their secondary differ-ences. There is still confusion; we have not yet entirely attained that condition of permanence which is essential to secure the full fruits of freedom. We are attaining stability, the dynamic stability of a ship in a seaway, not the static stability of a rock in a hollow. With it we will acquire the flexibility of a growing organism, not the frozen aspect of a design. Under it we can develop the full strength of free men.

From the weakness of our early years as a nation, we have advanced toward that full development. We have learned some-thing at least of how to preserve the full vigor of the democratic system against the control of specially powerful interests, as the present political turmoil demonstrates. We have learned how to make it possible for great enterprises to be carried out, covering an entire country, without allowing them to threaten the power of government or to victimize the people; we have preserved the advantages of competition without throwing everything into the stifling atmosphere of governmentally controlled cartels in order to avoid unbridled combinations.

We have gone a long way toward the organization of labor for the full assertion of its part in the national life, with a clear indication that we do not propose that there shall be dictation to government by those who may seize personal control of such aggregations. We have preserved the opportunity for individuals to create great commercial aggregations and new products, opening wide new areas of opportunity and profit, while at the same time we have introduced, and accepted generally, a system of taxation that securely prevents the creation of a permanent hereditary aristocracy of wealth. The principles of democracy have been carried a long way into industrial organization and into the organization of labor. There are autocratic industries, a few of them, and there are labor unions that are essentially dictatorships maintained by political intrigue, but these are exceptions. Industry is not autocratic, nor is labor organization, except in spots. The essentials of democracy are present: responsibility of those who rule to a complex constituency, freedom of discussion, and most notably freedom for criticism of those higher up the ladder. Any foreman or president or labor leader anywhere can be removed by those above him; in a democracy, we can be sure also that he will disappear in time if those below him are thoroughly discontented with his performance, even if he may sometimes survive for a while by family influence, or cunning, or sheer dishonesty and hypocrisy. Democracy operates in our armed services. They are supposed to be autocratic in the extreme; yet all who have worked with the system know that in spite of entrenched and vested interest no major or commander can last forever unless he measures up in the eyes of his men, and that those who artificially persist and perform crudely are regarded, both by those above and below, as anachronisms that should not be allowed in a well-working organization.

We have the encouraging spectacle of men of means leaving position and sacrificing income to enter public service without hope of reward or recognition, to enter it even at times in spite of expected gibes, men with public spirit actuated only by a desire to serve the public interest in time of need—in time of war and even in times of stressed peace. This country has produced, too, an extraordinary number of men of wealth who have regarded their accumulations as a public trust and have utilized them intelligently for the public benefit, often furthering good causes that could be furthered in no other way. We have a country in which other generosity is widespread, where alumni come to the support of the schools that launched them, where community chests flourish, where great numbers of the population seek out means through which they can express and implement their underlying altruism in concrete ways large and small. We have men of large affairs and a professional spirit who manage their enterprises in balanced fashion for the benefit of stockholders, customers, and labor, and who take pride in accomplishing this well and would do so even if there were no other great rewards. We have the salutary and stabilizing influence of labor leaders who rise above the lure of the advantages of belligerency in climbing the ladder of the labor movement, who lead their groups into sober participation and collaboration rather than deliberately into strife, that they may secure a just fraction of the fruits of production and preserve the health and prosperity of the organization that produces. These men of every stratum who thus serve their fellows for something other than immediate gain are exceptions; yes, they are exceptions to the general smallness and selfishness all about us, but they are not few, and while they are present the scene is not altogether dispiriting.

Without creating a permanent pauper class, we have gone a long distance toward a system under which no man who is not completely antisocial can fall too low or be too completely in distress. We have eliminated the labor of little children, and to a considerable extent we have overcome the modern form of serfdom in which men work in dangerous or disagreeable ways merely because they cannot leave.

We have not abolished all fear, but we have certainly removed a lot of false idols from their pedestals, and we have shown in strange ways a recognition, and even a respect at times, for sound accomplishment that is not measured by mere accumulation of wealth or raucous ballyhoo but is soundly assessed.

We have learned, and this is important, to carry on our political activities without becoming so excited as to lose all reason, to conduct them in fact almost good-naturedly and without rancor when we are at our best. We have produced an atmosphere, appreciation, and environment in which science and its applications are proceeding more effectively and rapidly than anywhere else on earth. We have produced, by far, the highest material standard of living that the world has ever seen.

In spite of our remaining bigotry we have created a tolerance and understanding under which great questions can be soberly and dispassionately examined in full, by all the people who wish to participate, to an extent that goes far beyond anything this confused world has ever before witnessed. We have produced a national power and consciousness, which is still in embryo, it is true, but which has at times produced national acts and foreign policies based on more than expediency, based even in some cases upon a desire to be mature and fair and to stand thus in our great strength in the eyes of the world, without having simultaneously embraced any philosophy of empire or conquest.

Certainly, without full definition and without accurate focus, we have attained understanding of the threat of totalitarianism and can act on it. This has been recently proved. We have fought two great wars and fought them well. In spite of all the clamor, we did fight to preserve our way of life. Confusion, disillusionment, were present, and the great songs were often lacking, but we fought with cold, unemotional conviction the last time; and we are still free, although the threat was real and dangerous. We demonstrated not only that a free people could fight well; we demonstrated that they could be far more effective in all the complexity, waste, and emotional drag of modern war than any regimented people, as those qualities of resourcefulness and initiative, matured best in true democracy under freedom, girded themselves for grim business. Moreover, we came out of the holocaust with our heads high, and have tackled well thus far the disheartening task of helping to reorder a sadly shaken world, not only for our own countrymen but also for many of those free people who suffered sorely and directly by conquerors or bombs. All this did not happen by chance. The spirit of freedom was alert. And it was alert by reason of the many who somehow sensed the occasion, and who spoke to their neighbors in the field or in the shop.

Now, because we are apt to overemphasize the rocks in the stream that impede progress and fail to appreciate the great current that flows between, this catalogue of accomplishment has been emphasized. It reflects our general pattern, in which everyone contributes more or less to approved procedures and the correction of mistakes. That pattern is an honest one, and it demands that we look searchingly at flaws as well as virtues. We shall do so. We have by no means accomplished all we should or could. There are plenty of places in which we can still criticize, faults that are only too obvious, acts at which we blush.

Democracy is not always strong. It can be very weak indeed, both at resisting external aggression and at maintaining internal stability and progress. Democracy is flexible and its keynote is freedom, and that means that it can be altered to be whatever the citizens freely desire. If they wish to have a wild scramble for petty advantage, and lose sight of their national interests in the process, they can have just that. If they turn their liberty into license they will not long enjoy either.

The stress is intense right now of course, for many reasons. First, there is the inevitable fear and confusion of the postwar readjustment, with much of the world in distress, property destroyed, and more people in the world as a whole than can live decently on the shattered means of production. Second, there is the recognition, for the first time complete, of the organized power that resides in those elements which would enhance the chaos in order to construct their own system out of the wreckage, and a recognition of the potency of the appeal of this philosophy of despair to those who see no light. Third, there is the advent of the atomic bomb, and the vague rumblings of some terrible form of biological warfare, and the intense fear that science has outreached itself, and that man now, suddenly endowed with powers far beyond his wisdom, will commit suicide. The first two are bad enough—alone they would have produced dismay—but the addition of the third comes near to being just too much for a world shaken by the terrors of the most awful war in history, appalled at its own depravity as exemplified at Dachau and in numerous other places, and facing up to a future that still contains nearly all the ignorance, prejudice, rapaciousness, and cupidity that have plagued it since the first simian climbed down out of his tree and started the adventure. It is no wonder that the shock has produced a paralysis and that the optimistic note is now only too likely to sound hollow and artificial.

The central question is whether there is order and control, whether orders are carried out promptly and cheerfully, in the government itself and in the organizations that are subordinate to it. The usual word in this connection is discipline, but that has a connotation of conformity by reason of fear of punishment, whereas one of the primary objectives of the democratic scheme of organization should be to remove fear as a motive from all save the incorrigibles, who presumably will not function without it. Here is the rub. Is it possible to have an organization, say a business concern, from which fear is essentially absent that still will work? Can such an organization allow free discussion of policy questions, at the level of primary concern, criticism that proceeds both up and down the ladder, officers of every rank

dependent upon the consent of those under them for their continuance, and still maintain the system of control that is essential if it is to be more than a debating society? We know it can
sometimes be done, for we have all seen the leader of such ability that his men would follow him anywhere and execute his
every order with alacrity simply because of confidence in him
engendered by just such discussion. There is a vast difference
between the atmosphere in the average industrial concern of
small size today and what it was at its worst fifty years ago,
and we are making progress toward an ideal. Yet we know full
well that if we went to extremes at once we would have confusion and not much else. We still have to have a good deal of old-
fashioned discipline to get things done.

If democracy is to be at its best this difficulty has to be solved.
Someday we may be able to go a long way toward a system in
which voluntary conformity to group opinion is fully controlling.
In the meantime we need to be sure that, just because policies
are debated out loud, just because we have free speech, we do
not create soft or wishy-washy organizations as a result, where
free expression of opinion before the event gives place to malin-
gering after the decision is made. We also need to be sure that
we do not confuse democracy with government by committee.
Most of the important actions, the executive acts, need a man
with authority who can make decisions and carry them out. Democracy can place him there; it can see that he listens to advice
and remove him if he does not, but it had better back him when
he makes a decision and see to it that his subordinates jump
when he speaks.

The progress of the United States Army is pertinent in this
connection. Because of our inherent dislike as a people for military discipline, because of our standard caricatures of the drill
sergeant, there is no doubt that, relatively, we have the most
democratic army in the world. Yet its performance in battle has
clearly shown that it did not lack discipline. In fact, the resourcefulness and initiative of the individual soldier, made possible by
the fact that discipline is reasonable and not a rigid thing, made

much of the magnificent performance of that army possible.

We shall learn to do better as we learn more about democracy. There has been altogether too much emphasis at times on the privileges of citizenship. It is time we emphasized more of the responsibilities that go with freedom. The duties of the citizen include following orders, when given by constituted democratic authority, just as promptly and thoroughly as though they were issued by a martinet in a dictatorship backed by the threat of the concentration camp. When they are followed in a democracy, they are likely to be followed with intelligence and vigor; that is one of the cardinal virtues that can make democracy great and strong. Better application of it will help overcome some of the difficulties that hamper us.

A system of free enterprise has its parasites and has not yet learned how to get rid of them, any more than a dog has learned how to rid itself of fleas instead of only biting one occasionally. They are static in the broadcast to all those who have to listen to them, especially when the successful tricking of neighbors is followed by ostentation before those who seriously labor. It is difficult for a free society to learn how to get rid of them, because of its very freedom. Public opinion alone would do much of it if parasitism could be surely defined, but even this is difficult, for there is a gradual shading from parasitism to beneficial but unusual performance in the complicated ebb and flow of modern commercial activity and the market place. Moreover, no one objects to play, or to plain and fancy loafing, for that matter, when it is earned, and it is often very hard to tell whether a given man is rowing his full weight in the boat or the weight of ten men, or just taking a ride. But parasitism is not unique to the democratic system or even to the system of free enterprise. Hitler's war machine was crowded with parasites, usually all decked out in resounding titles and decorations. There has been plenty of parasitism in every social order that has ever been constructed, from the first colonies of ants, and the Roman Empire, down to the present.

We have not yet learned to keep a free-enterprise system on

an even keel. The business cycle haunts us, especially as our adversaries cynically bide their time for the next bust. We have not learned how to keep employment full at high wages, with reasonable and attractive rewards for those of courage who venture their possessions well, without getting tangled in a spiral of inflation or encountering painful and sudden readjustments. We have not conceived a system that will ensure that those who labor secure their reasonable share of the product, a system that will fully preserve freedom, without going through strikes and readjustments that cause grief and distress all around. We have not escaped from a net that would fix the pay of antagonistic groups in accordance with their opportunity to disrupt rather than with their contribution to general welfare. We have not yet come to the point where every youngster with real talents and energy, wherever he may arise, is assured of all the educational opportunity that he can effectively encompass for the ultimate good of himself and his fellows. We have not yet learned how the course of government can best be directed in the interests of the great mass of the inarticulate, in the presence of highly vocal, specially organized interests. Nor have we completely eliminated skulduggery from our legislative halls or anywhere else.

Our present dangers are clear as a bell before us. One is that we may not be able to continue our highly successful system in prosperous operation while we spend heavily for armament and for the rehabilitation of our friends. The immediate problem is that of maintaining a safe balance and remaining on prosperous middle ground between a spiral of inflation and the sort of full-fledged depression that nearly wrecked us not so long ago. The absurd shortsightedness of some great industrial units relying on scarcity and producing undue profits has furnished one pressure toward inflation, and the seizure of enormous power by labor leaders who forced round after round of wage increases has furnished another. We cannot afford to allow industry-wide combinations, either of managers or labor leaders, to exact undue prices or wreck an industry at the expense of the public. There were exceptions, of course, leaders of industry and leaders of labor with statesmanship and vision. But too many failed to grasp the national danger in its imminency and played an immediate selfish game. There were too few thinking of the overall public interest, as we juggled with the problem of proper sharing in the product by those who produce it: laborers, engineers, managers, people who have saved and invested. The result a while ago was an upward spiral that has now ceased to climb, and we face the problem of settling to a reasonable level without entering upon a spiral that descends. It will require statesmanship, more than we have had.

A less immediate but more nearly fundamental problem is to prevent special interests from raiding the treasury, unbalancing our budget, and destroying the national credit upon which the whole world depends. There is no use here in aiming the finger of scorn at any one group; many have been at the trough, and many more are honest advocates of genuinely good causes. In most cases there is an underlying and sound need and reason, whether to maintain a strong merchant marine, or to give the independent farmer a break in a tightly organized world, or to ensure that a veteran is reimbursed and not merely forced to serve his country at low wages in the presence of high ones. There are great underlying matters of policy, such as the necessity for preserving our diminishing soil against erosion or of its minerals against selfish mining, or to manage our forests for posterity instead of baldly destroying them for a quick profit. But no one will deny that there are also powerful interests which would merely enrich themselves from the public purse or by artificial influence or regulation mulct the public.

The problem is thus a dual one: first, to separate the merely selfish from those measures which are actually in the broad public interest, and, second, to balance these one against another and to restrict the total effort to a scope that can be afforded, so that as a country we live within our means, distribute our product wisely, and do not attempt to distribute more than we produce. All this calls for patriotism and judgment of a high order in our legislative halls, and calls for it in circumstances where those who legislate are harassed and badgered, often with devilish ingenuity, by those who clamor for this or that. It is a wonder we have done as well as we have, thus far.

This is a central problem as we look forward. It is summarized in the idea of the welfare state, the state, in other words, that is deeply concerned with the welfare of its citizens and that uses its power directly to provide them with relief in distress and services to increase their health and comfort. England has gone further down this path than have we as yet, and is in serious difficulty, but we are moving down it steadily. As in most such problems there is a reasonable middle road, but it is slippery and it is hard to drive on the crown and not slide into the gutter. Neither extreme is attractive. We certainly would not return to a system in which the unfortunate would be entirely dependent on private charity; our basic concept of the dignity of man forbids it, whether the misfortune be owing to age or disease or industrial vicissitudes. On the other hand, the opposite extreme is equally unattractive in a world where the maintenance of our full strength is a must. In the extreme, we have a horde of bureaucrats extracting heavy taxes from all who are at all prosperous in order to furnish paternalistic care and control to all who are less so. They take two dollars from Jones to furnish one to Smith and make Smith stand in line to get it. Carried to extremes, this system means that no private wealth of any magnitude can be secured or kept, and all new business, or old business, for that matter, becomes a governmental affair, without the sparkle of private venture. Carried to extremes, also, it can create a class of loafers supported by taxation, for, unfortunately, the love of work is not universal, and there are many who would bask in the sun and scoff at those who produce, even on a pittance. It is no way in which to progress and maintain strength in a world in which technical innovation is rapid and necessary, and we know it, but we nevertheless drift toward it because each new humanitarian project appeals strongly, and we do not stop to count costs or compute taxes. The insidious aspect of the matter is that benefits can be paid, for a time, out of capital rather than income—that is, by seizing private savings or extending national credit. This is justified, in fact essential, in war and in other dire emergency. It takes resolution and understanding to set limits as the emergency passes. When those limits are set by the exhaustion of the capital itself, by getting to the limit on taxation or credit, a sorry day of reckoning results. It remains to be seen whether this country can balance its affairs and limit governmental action to where it is really necessary and efficient, before we crowd ourselves with burdens that deaden the whole and find we have taken care of everyone at the expense of everyone else and failed to take care of the primary national interest. We cannot afford, today, to interfere unduly, even in the name of humanitarianism, with the diversified vigorous private initiative that made us great.

A third and related hazard is that of throttling new enterprise by shutting off the source of venture capital, and enmeshing in the red tape of bureaucracy the new small pioneering concerns that should be encouraged to build the great industry of the future. We are coming close to driving ourselves down the path that England has started to follow, and on which it now hesitates, the path to state socialism, even though there is no doubt whatever that the people of this country would have none of it, were they to face the issue squarely. We have a tendency, we, our legislatures and our courts, to take a crack at anything large or prosperous, and many of the blows are badly aimed and hit the little fellow. We have even come to look askance at success itself, until we are in danger of destroying the very thing that made this country prosper.

We go about it in subtle ways, and as a nation almost subconsciously. Again there is a sound basis for much of what we do, but we fail to grasp fully the overall effect. We regulate the security exchanges, and they certainly need it, but in the process we inevitably frown on the unconventional. We tax corporations and individuals, and we need the money, but we tax in a manner that stacks the cards against the pioneer who would take an industrial risk. We regulate the conditions of labor in industry, and we can certainly not have too much health or safety, but we forget that in the process we may force the employer of ten men to spend all of his time filling out forms. We proceed against monopolistic use of patents, and patents should not be used to circumvent the antitrust acts any more than should any other property, but in the process we weaken the system that is the main reliance of small concerns attempting to establish themselves in the vicinity of big ones. We know we do not wish to tolerate unregulated monopoly in business. We realize that there is a tendency for business to aggregate, that this may involve economies and public benefits, but that if it goes too far it brings in abuses we will not tolerate. In a vague way we know that the principal offset to this tendency, outside of the regulated utilities, is the continuous advent of small pioneering industrial units. Yet our acts are usually negative and oppressive, and they often stifle the very thing we would cultivate. When we turn to positive measures, we substitute government paternalism for direct encouragement of freedom of action on the part of the pioneer. If we do not watch our step the end is inevitable and drab. Above all, we need to return to the conviction that he who makes two blades of grass grow where one grew before, who creates a new product or service that the public desires, who is creating worthwhile things that are genuinely new, and following the laws of the land and of decency as he does so, is a public benefactor. We need also the conviction that the best way to encourage him is to unwind the red tape from about his neck and allow him to keep a reasonable fraction of his gains. This attitude will not be reflected in legislation until it is prominently present in the public mind. It will not reappear until we understand, better than we now do, the true meaning of the word security in a hazardous world.

Why point out these dangers? This is no book on economics or principles of government. It is concerned with the interplay of science and democracy and war. It attempts to review and look forward, to estimate whether we can avoid another great war, and, if we cannot, what its nature would be and how we could prevail and preserve our way of life. But the heart of this matter is that we must preserve our strength, and our strength resides not merely in our planes and fighting ships, not merely, in fact, in the success we may have in applying science to war— it resides in the whole strength of our democracy, political, economic, and moral.

If we lose our strength we cannot meet the threat. If we fail to solve our problems we cannot be strong.

There is a tendency to regard all this as apart from our daily affairs, a matter for presidents and legislatures and judges. It is not. Every little act we do, every opinion we express, affects the outcome. The question of whether a sound citizen or a persuasive crook becomes sheriff of the county does not affect alone the effectiveness and honesty with which local affairs are conducted, it determines whether a part of our democracy is strong or weak, and many such acts make up the whole. When the humble citizen votes, or sits on a jury, or discusses affairs with his neighbor at the corner garage, he may little grasp the connection, but he is, for his fraction of the summation, determining whether the bright youngster down the street will have to die on some future battlefield, and if he does whether he will die in vain. The great measures are determining of progress, but they are founded on all the little ones, and they are responsive to the will of the people. If that will is that we shall be ready to fight if necessary, and by this means avoid the necessity of so doing if it is at all possible; if there is understanding of the relation of this to the strength of democracy, and that that strength depends upon the every act of every citizen; if the will is strong enough to rise above petty motives or selfish prejudice, then the outcome need not be in doubt. Democracy will flourish, and we shall be strong.

For our encouragement, as we examine the current scene about us and are sometimes dismayed at the difficulties we face, we can look back on the performance of the last decade. In war and peace we have accomplished much that no dictatorship could ever match. We fought a war well and applied science in the process in ways that startled the world. When peace returned we created an internal vigor and prosperity that enabled us to lend a helping hand to those, our friends, who suffered more grievously, as they started their climb back to national health. We have not come to the end by any means. The applications of science yet to come are manifold and far-reaching. With them we can establish a standard of living in this country far higher than we have ever had; we can make more goods and have them more generally available throughout the population. We can prolong our lives and escape the ravages of old age, overcoming the scourges of mankind, epidemic disease, cancer, senility, to an extent that we can now barely grasp. We can create an environment in which the creative arts can flourish, in which the human spirit has an opportunity to rise and aspire. We can build a society in which there will be justice and good will. All this is within our grasp; we know it, for the performance of the past ten years is a guarantee of the effectiveness of the system under which we operate and of the fundamental principles to which we adhere. All we have to do to bring it about is to preserve that system and improve it and hold fast to those ideals and the faith from which they arise.

By no means all of the problems we face as a democracy are here reviewed; they cannot be. Some have merely been mentioned, and others stated in condensed form. There are two matters in which the operations of science in a democracy are especially prominent, one having to do with education and the other with how we plan our military program in the light of scientific advance. To these we now turn.

CHAPTER XVI

EDUCATION

"Anyone familiar with education knows that for a very considerable portion of the population it is the family financial status which places a ceiling on the educational ambitions of even the brilliant youth. The oft-repeated statement in certain smug circles that 'any boy who has what it takes can get all the education he wants in the U. S. A.' just is not so; it is contrary to the facts."　　　　　　　　　　—JAMES B. CONANT
Education in a Divided World. 1948

OUR CONCERN in this chapter, as throughout the book, is with the workings of science in a democracy and the relation of this to national strength. In considering educational problems, we come directly to the conclusion that there are certain special types and means of education to which we need to pay special attention now that there are atomic bombs and the possibility of biological warfare.

In our national thinking in regard to education there has always been confusion. We are prone to emphasize equality of educational exposure and to lose sight of equality of educational opportunity. Our thoughts become fixed on raising the general level of education, and we have not lent enough encouragement as yet to those who could rise well above this level and contribute mightily to the general welfare—and to the safety of the nation, as a result.

These two aspects of the subject are complementary—not conflicting. We need both, and it is a matter of emphasis. The logical balance of emphasis has shifted, now that the application of science to war has become a determining factor in our future, and now that full national strength in every phase of national life is essential for our national well-being and perhaps for our national existence, and we have not yet shifted our sights accordingly. There is no diminution of the need for a high level of general education of the people; there is the need for super-imposing upon this better facilities for seeking out special talent and giving opportunity for all the educational advantages it can well absorb.

The bulwark of democracy is education; this conviction is deeply imbedded in our national consciousness. The little red schoolhouse was the principal reliance of the republic, and we know it. Not only should every youngster have the opportunity to learn to read and write and figure; he should be forced to submit to a process of learning these things at public expense, based on taxation by which those without children should pay their share for the elementary teaching of the children of others. Moreover, this process touched the personal interests of the people so closely that our forebears early insisted that the sys-tem be divorced from the maelstrom of politics, and woe to the politician who tried to control the schools for patronage pur-poses. Moreover, while religious sects may create and support their own schools or colleges at their own expense, we have held throughout to the principle of the divorcing of church and state.

As time went on, the years of required schooling were ex-tended, but we adhered in general to the system and insisted upon it. Notably we have kept elementary and secondary public education a local affair, supported by local taxes, out of fear of central control, for we insisted that the town school be man-aged completely by the town fathers, and we watched the per-formance of these worthies with our eyes cocked for irregulari-ties, even when we tolerated them blithely in other town affairs. Graft, proselyting, manipulation of funds, and rackets prolifer-ated in many a local government, but the schools were generally left alone because the people insisted on it. Hence, as we now consider a great plan of Federal support of elementary educa-tion, which means that prosperous states should pay part of the costs for less prosperous ones, and which seems to most of us to make sense as a further guarantee of future prosperity and stability of the republic as a whole, we hesitate for the old reason: we have no intention of allowing our schools to be politically controlled, or controlled by a centralized bureaucracy, and therefore we search for ways of using Federal aid without such entanglement.

Beyond elementary education we have followed a dual path. On the one hand state universities were established, tax-supported and more or less successfully set aside from politics, although with scandalous failure at times in this respect. In general, at the start, these followed the philosophy of equality of educational exposure rather than equality of opportunity for educational advancement. Any son or daughter of a citizen could attend, without payment or on merely nominal fees; and many such institutions promptly became centers of mass mediocrity as a result. On the other hand, private institutions were founded in numbers long before the state university appeared, usually with a sectarian origin, from which the greater part of them later departed. Fees were charged, and in the early days these institutions were regarded primarily as places where the clergy could be trained, where the sons of merchants could acquire a bit of polish, and where gentlemen could appropriately gather. Only too often they became loafing places where social life was uppermost and true scholarship was considered not worthy of the gentleman. With notable exceptions, for true genius and scholarship will make itself felt even under the most depressing conditions, the whole university system of this country less than a century ago was mediocre and far behind the universities of Europe, so that the really ambitious and accomplished turned there for the education necessary for the professions. The difficulty was that the basic philosophy was wrong: on the one hand, dead-level education for all, and, on the other hand, higher education only for the privileged.

All this is now radically altered. As the states developed pride in the standing of their universities and supported graduate study and professors of eminence, many state universities rose to an accomplishment comparable to that of the finest private institutions or the finest that Europe could offer, and our whole system of higher education rose to levels that no longer needed apology. Men of great wealth poured endowments into private institutions, enabling many of them to support advanced programs of real scholarship widely spread over the sciences and humanities. The renowned professor was supported so that he could live well and remain independent of selfishly induced favor. The idea of academic freedom gained ground, until men could speak their true thoughts, or even absurd theories, or bids for notoriety, without fear of losing their professorships as a result. Most important, systems of scholarships and the like appeared, so that it became possible for youngsters of ambition and talent to go through to the most advanced degrees without parental support, with hardship and unwarranted burden and impediment along the way perhaps, but successfully and in a manner that toughened their fiber; and many a valuable citizen pursued the path. Even then, however, we had merely touched the borders of true equality of educational opportunity.

The educational structure of the United States today is an approximation, but only an approximation, of the system that Thomas Jefferson visualized in the early days of the republic. For the ultimate safety of the new state he saw two needs: first, for mass elementary education at public expense, and, second, for state-supported education of the highest order open without expense to those who could qualify for it in competition. He visualized an electorate educated throughout at least in the elements, and a system by which leaders could arise, through fair selection, educated to the maximum degree that they could absorb to advantage, supported by the state as they acquired broad background and advanced skills to be employed in the professions for the public benefit. He thus proposed a pyramidal system: elementary schools for all, high schools open to those selected by competitive examination, colleges for the few who could similarly qualify, and universities at the top of the pyramid, manned by scholars and attended by the youngsters most highly endowed from an intellectual standpoint, provided they had ambition and were not afraid of hard intellectual effort, who should be drawn from every stratum of society and supported quite independently of their personal or family resources. We have approached the ideal, after a struggle of more than a century and a half, but have by no means reached it.

Today there is a fresh and impelling necessity that we should do so. In a world where wars were crudely fought, with little relation to industry or the application of science, we could coast along fairly safely. In a world where the prosecution of war or the avoidance of war demands that we be in the forefront in the applications of science to public health, industry, and preparations for fighting effectively in a modern sense, we can no longer afford to drift with a slow current. It is essential that we provide equality of opportunity of higher education in the full sense, so that talent and intellectual ambition shall have no artificially imposed limitations, so that highly endowed youngsters, wherever located, may come forward with full educational equipment to attack the great problems of the future, in law, medicine, principles of government, social relationships, science, engineering, business theory and practice, and in the humanities that underlie all our thought on the problems of civilization. Moreover, we need to accomplish this without becoming enmeshed in political patronage or entrenched bureaucracy. Imperative as it is, to accomplish this will not be easy.

If in the present discussion we hold to the point of view maintained thus far in this book—the interrelations of science and democracy as regards war and survival in a troubled world—some clarification of the problem may be expected. From that point of view we can readily discern special areas in education where the need for advance and improvement is greatest.

One such is medical education.

Originally, in this country, the only medical education of any merit whatever was a system of apprenticeship, where the professional man of skill drew disciples about him, taught them what he knew, and launched them on their careers. The art of medicine was then severely hampered, for the science of medicine was almost entirely lacking. It was only because the race was prolific and rugged that it was not more hampered by its ills in these early days. The profession had its quota of men of character, more of a quota than the other professions undoubtedly because of the heavy direct and personal responsibilities involved, devoted and intelligent men who took the oath of Hippocrates seriously and dedicated their lives to the good of their fellow men, without undue modesty or timidity, standing upright in their pride of profession and their dignity of ministry, bringing light and decency and professonal competence to an otherwise faulty and at times sordid system. But, even though tempered by the presence of highly qualified and devoted men, the old apprentice system was bad indeed.

All this is now changed. The rise of the modern high-caliber medical school, pioneered by Welch, Osler, and their colleagues at Johns Hopkins, the influential survey and prescription of Abraham Flexner, the gradual development of general public understanding of the difference between the charlatan and the educated medical practitioner, have combined to give us a great system of sound medical schools, with high standards of admission and performance, alliances between medical schools and hospitals, systems of internships and residencies, and in general a structure of medical education that is effective for its purpose as far as it goes. But this great medical-school system has two faults: it does not train enough doctors, and it is a far cry from our ideal of equality of educational opportunity.

Part of the reason for these difficulties is that the medical profession in organization and philosophy still recalls the guild system of the Middle Ages, resisting control from without, holding to a system of teaching by apprenticeship and a long, severe path to mastership. The career of teacher, here even more than in other professions, has not been made sufficiently attractive, compared to other professional careers, in income or recognition, to attract sufficient teachers of high caliber. For this reason, and because the educational process has been made so long and arduous that only youngsters with financial backing can hope to enter at all, and only those from families of considerable means can hope to pursue the extended and rigid path to the specialties, there are not enough good doctors to go around.

This last is serious. When it often takes twice the time, and more often twice the investment, to go through the educational mill of specialized medicine, compared to what it takes for corresponding preparation for law, chemistry, physics, or biology, with these latter professions rapidly expanding and becoming daily more attractive, medicine is just not going to get a good selection of youth for its purposes. Moreover, young men who finally emerge from the treadmill are not likely to do much research; such spark as they might have had was only too often deadened long before, and they will proceed to recoup their utterly depleted finances if they can. There are exceptions, of course; it is almost impossible to suppress true genius or the intense ambition to create, and so notable figures continue to arise, but the system as a whole is not designed to bring them forward.

Medical men realize this difficulty and seek a way out. All that can accomplish it rapidly is Federal money, and plenty of it, and this involves the risk of bureaucratic control that would deaden the system further. One possible solution is to provide Federal scholarships on a broad and generous basis, assigned through a competitive selection process managed by representative boards of citizens, not by medical men alone, with the subsidized youngsters free to choose their own schools and pay generous tuition fees, thus placing the schools in a genuine competitive position. Even this may be accompanied by more control of medical affairs by laymen, for when public funds are used the customary checks will be invoked, but lay control and bureaucratic control are not synonymous. The medical profession will be wise if it aids in steering this undertaking into sound channels. This is a move toward genuine equality of educational opportunity in a field where it could produce prompt and very salutary results.

Why is it that there is a new and imperative reason for real progress on this matter? We have briefly discussed the possibility of biological warfare. The indication is strong that at present it is not of major importance in connection with a war that might occur soon, rather that it is an auxiliary method, with a strong possibility that it would not be used at all, and serious primarily in subversive warfare. But there is no telling what a generation may bring. Methods might indeed be developed that would make biological warfare a major threat, comparable in its scope to the atomic bomb and far more terrible and repelling in some ways. Or it might be that defensive methods, based on exact knowledge of the underlying science and intelligently applied to the conquest of disease, to public health, to sanitation and the protection of foods, to more powerful hormones, antibiotics, and the like, would hold the fort and force biological-warfare methods to continue to be secondary in potentiality. The alternatives are critically important to all of us, and so we insist that there be plenty of basic scientific research and plenty of applied science in this whole field, and of the highest possible quality. For this, sound and widespread educational opportunity is essential.

The great advances in medical science have not originated in the past to a sufficient extent in the research of medical men themselves. They have come more often from the chemists, the bacteriologists, the physiologists, in the universities and in the pharmaceutical industry. They have arisen as by-products from such unexpected places as the dye industry. With notable exceptions, of brilliant medical men who have created an entirely novel approach to a medical problem, the advance of medical science has come about because other and neighboring sciences were progressing at a prodigious rate, and applications were bound to occur. The primary contribution of medical men has usually been in clinical research, which has often been merely the clinical testing of semiperfected means, or in research in surgery or the like, utilizing products and instruments that appeared outside the profession and aimed originally at other purposes. There is a gulf between medicine and the other professions, widened at times by the insistence of the medical man upon his prerogatives, and crossed only too rarely by medical scientists of true breadth and vision. Yet, if we build the biological, chemical, and physical sciences strongly enough, if we have strong universities and industry where they are applied with versatility, continuing advance in medicine can be expected. We can, in time and in these circumstances, expect the gulf to narrow. In a quiet world, with a quiet future in prospect, that would be enough. For the present situation, it is not enough.

This is why we must make it possible for any youngster anywhere, who has unusual and especially pertinent qualifications, to proceed to the doctorate of medicine at public expense and to engage in postdoctorate research, in direct medical fields or in any of the neighboring fields, on a dignified basis where he will have enough to live on reasonably, without being dependent on auxiliary patronage, without postponing his marriage unduly, and free to seek out the masters of his choice. Thus we can produce a strong cadre of highly trained men on whom we can rely for the advances that may well save us from disaster. The entrance should be one of rigorous examination, repeated as the steps in academic advancement occur, based on the best examinational procedure that can be devised by experts in that field, extending far beyond the ability to answer trick questions, and aimed at selecting those who are truly gifted in extraordinary fashion. The opportunity to compete should be open—not available through tortuous local political paths, but open—to every youngster of promise and character in the entire population. Even if it is rigorous indeed there will be plenty of talent, plenty of genius, for that matter, to man the profession.

With this needs to come a revision of the educational process in medicine, carried forward by the broadest and wisest individuals in the profession, not left to the secretary of some committee or argued in chaotic form in a Congressional hearing. The best of the profession are overburdened, as an inevitable result of the general lack of numbers, yet they alone can do the job. In fact, strong groups are already seriously at it. Perhaps there is another Flexner in the offing somewhere to take the lead. To a layman there seem to be plenty of opportunities to head in on the problem. Why, for example, require a full college course for entrance to the best medical schools when the curriculum thereafter excludes the humanities, young doctors in the groove have almost no opportunity whatever to broaden their education, and the start toward wider culture is almost inevitably sacrificed? The engineering profession, fully as concerned with the practical application of science, is learning rapidly that an engineer must be a strong citizen and that for this he needs contact with the humanities throughout his educational career, if he is to be expected to progress broadly thereafter. Does it make sense to open medical education with a severe test of memory? The chemists learned long ago that the effect of this was to repel and deaden, and they have just as many facts that they use but are content to keep in handbooks to store the nonessential and the trivial, freeing the mind to grasp basic essentials. They have just as tough a job in teaching the necessity of precision of statement or of plain hard work. Is it reasonable, today, to insist on leaving in the curriculum all that was introduced there years ago and merely adding everything under the sun? Is there no basis in medicine for separating the essentials from the trimmings? Is not medicine, par excellence, a field for the wide use of the case system, which has thus far been introduced timidly? The profession of law learned long ago that to try to teach all the basic law a practitioner needs would be hopeless, and they have saved their situation by the case system. Earnest medical groups are attacking these and allied questions, and if they can come up with sound answers they will merit strong support.

We need more doctors and more broadly trained doctors. Moreover, we need to see them more often take positions of responsibility in the profession when young and vigorous, to avoid the static paralysis only too likely to creep upon any organization controlled by old men if the youngsters are too fully in dependence or subjection, as is far too often the case at present. Medicine, along with the other professions, is now taking wider responsibilities in civic and political life, dealing more directly with our future health and security as a people, and a departure from intense specialization must somehow find its place in the sun of medical approval. But there seems to be another reason for breadth of vision and experience which is peculiar to the medical man. All professions deal with human life and well-being indirectly. The medical man deals directly with his patient in a highly personal way. This is not a matter of test tubes or microscopes; it is a matter of human understanding. Many facets of human relationships are absent or distorted in the artificial atmosphere of the hospital or the consulting office. The doctor, above all professional men, needs to be a full man. For this his education needs to be removed from the rut it has long traveled and given the breadth that the intimate responsibility of the profession needs for its best service to those who depend upon it. The finest of the old-time country doctors had something rare and subtle that was without price. Part of it came from the ideals and experience of the profession itself. Much of it came because they were, throughout their careers, valued and respected members of the communities in which they lived, learning life from the wisest of their fellows and teaching it in their turn. We can recapture some of this without sacrificing anything of moment in scientific rigor or detail.

Medical education thus gives us an excellent example as we look to the future. If that future may involve biological warfare, then we most certainly must have an adequate and highly talented medical profession, for the greatest safeguard against such methods would be an advanced standard of public health. Yet, even from the limited standpoint of the risk of the use of biological methods in war, this is by no means enough. Other professions are involved, and we shall digress to illustrate the point.

During the war just ended, together with the British we made spectacular progress in the wholly new science and technique of radar. We left the Nazis and the Japanese at the post in this

essential field. As a result, almost as a direct result depending
uniquely on this advance, we overwhelmed the Japanese Navy,
after we grasped the power of the new techniques and really
used them. We pounded the Nazis from the air, night or day,
and England protected itself against similarly continued bombardment.

The real reason we made such great progress was not bright
inventors or clever gadgets. It was the fact that we had thousands of men who understood the underlying science in the
field, who skillfully practiced the necessary techniques, who
were good gadgeteers. They were in our universities, throughout industry, and in all sorts of queer places in the general population. Enough of them were gathered together and saved
from senseless expenditure in tasks far removed from their skills
to do the job, all the way from the research laboratory through
pilot manufacture and engineering design to mass manufacture
and skillful use in the field. We made great progress because
we had the background for it. This had appeared because of
our science of radiation, because of our radio industry, and
because of our amateurs. It existed because for a generation
youngsters had pioneered in the field in many ways, advancing the boundaries of fundamental knowledge in the universities and the detailed techniques in amateur efforts throughout
the land, because we had large industrial radio laboratories
filled with able scientists and engineers, because we had an
intensely competitive manufacturing industry in radio, a hel-
ter-skelter affair, with units trying absurd things and going
broke, but with keen alertness to build gadgetry that would
sell along any interesting line that showed up. A regimented
or governmentally operated system of universities and industry would not in a thousand years have produced the background for such an abrupt advance. Neither would it have
occurred if our system of education, formal and informal, had
not produced an adequate crop of youngsters skilled in the
whole affair, all the way from the underlying abstruse mathematics to skill in nursing a balky thermionic tube.

Now, for the kind of world with which we have to deal, we had better be about creating a similar situation in the biological sciences. It will not create itself this time, for there is not the same popular appeal and opportunity for inconsequential application. Strong sources of independent support can be drawn upon. The progress of the pharmaceutical industry is involved in this need, and of other industries dependent on biology, such as the fermentation industry. There is a clear connection between biology and chemistry, and the magnitude and virility of our chemical engineering and our chemical industries are important factors. As long as these are in good shape, prosperous, forward-looking, and genuinely competitive, we have the basis for progress. These alone, and the fascination of the biological sciences themselves from an intellectual standpoint, will accomplish much. But all this will need to be supplemented.

The biological and medical sciences are closely interrelated and cross-fertilize one another. If the biological sciences advance rapidly on every front, so will the medical sciences. The opportunity is fully present. There is no more fascinating field, especially to a keen youngster, than the science of life in all its breadth; there is no area of intellectual opportunity more absorbing. It is opening up daily. Every few weeks bring the news of some new antibiotic—some chemical obtained in strange ways as was penicillin, which has the important property of stopping pathological organisms in their tracks. The chemistry of proteins, vitamins, hormones, and enzymes, and their newly appearing relationships, is just now unfolding not only ways in which we can live more securely in the face of disease, but also ways in which crops and animals may be molded and controlled for our use. The science of genetics enables us to alter species at our will; it will enable us to create new species for our benefit. It opens broad intellectual vistas of the progress of life on the earth and enables us to look into the past and sort out its evolution, just as astronomy enables us to look into the far reaches of space.

All this would be fascinating to a youngster, if he knew about it. We can probably never create the same atmosphere that obtained in radio—where there was an amateur expert in every block. But we can certainly create in this country a sound, broad, biological science with a host of professional and amateur adherents. To do so is to do much. It is to open up to many men intellectual opportunities of the highest order; it is also to build the soundest bulwark we now know how to build against the possibilities of being behind if there ever came an intense race in biological methods of making war. To accomplish all this we need first to clear the way for the really talented youngster to go to the top in the profession, there to become the research man, the teacher of others, the leader in new industries. We need also to be sure the profession is an attractive one for the humble worker therein, and it will be if enough talent heads into it and makes it so.

The medical and biological sciences have furnished a convenient example to illustrate the needs for enhanced educational opportunity for the gifted. They do not, of course, illustrate all of it. Just because we have made great recent strides in the applications of physical science is no reason we should rest on our oars. Chemistry, with new materials and its striking industrial advance, offers very great opportunity indeed. More broadly, all of the professions are essential to our strength and progress as a nation. Someday they should all be led by the most highly qualified individuals in the entire population, regardless of personal circumstances, furnished at public expense with all the educational opportunity they can usefully absorb.

There is also another way in which we can advance science, and with it scientific education, and we are pursuing it. This is by supporting research in the universities, for this sort of research in particular has as an important by-product the training of men. Since the war there has been real progress in this regard. It needs to be continued and steered into sound channels.

Fellowships for advanced study have been extended by the Atomic Energy Commission, the Navy Department, and the

Public Health Service. Advanced research under contract with the universities has been furthered by all three, on the whole wisely and soundly. This is an excellent beginning. But it needs to be smoothed into form, and extended, before it is cut short by budget pressure or embedded in a faulty system. The Navy, in particular, has done a magnificent piece of work in the field. It understands scientific war and university-service relationships and has managed the undertaking well. Yet, for the long pull, it is anomalous to have the support of an essential part of our university system reside in an armed service. It is dangerous to have it reside anywhere where it could come under the arbitrary control of a single individual, no matter how many advisory committees there might be. This is a hazardous thing that we do, the giving away of public monies, and we need to surround it by safeguards for the many years ahead. With the Federal government plunging into the support of research on an enormous scale there is danger of the encouragement of mediocrity and grandiose projects, discouragement of individual genius, and hardening of administrative consciences in the universities. Some of this we have had; it is much to the credit of those that have managed the affair since the war that we have not had more, in a system still essentially bureaucratic in nature. We need to centralize the effort inside the Federal government, and to place ultimate control of policy in the hands of a representative body of citizens, selected and confirmed with care, bound to justify their program annually to the Congress in order to secure funds, and supplied with a well-paid, competent executive to manage their contracting and business affairs generally within the framework of government and subject to all its checks and controls.

We have just such a system provided for in the National Science Foundation bill, now before Congress, long overdue. It has been held up for years by incidents rather than fundamental disagreement. Now it appears that there is nearly unanimous approval, and when the wheels of Congress finally revolve we will have a National Science Foundation. As it proceeds, if it is wisely supported, it can ensure Federal support of university research; it can provide fellowships for the brilliant; it can go a long way toward providing that equality of higher educational opportunity which we need to superimpose upon our educational system as a whole, in order to adapt it for our true purposes in this world of threats. It can formulate and support a sound governmental attitude toward science, and scientific education, in these days in which the burden of both has increased to the point where it can be carried only at Federal expense. It can guard against the heavy hand of bureaucracy and furnish a bulwark against political pressure on this vital aspect of our interests. It can further science, free science pursuing its independent way to unravel the mysteries of existence, carried on by free men whose guide is truth and whose faith is that it is good to know. As it does so, it can aid much to protect us all from the vicissitudes of nature and of selfish man.

This effort to strengthen education as the bulwark of democracy can be made in many fields besides those that may have direct application to war, but we had better begin with the ones where the need and opportunity are obvious, learn the methods, and expand gradually and deliberately. The first steps can be taken at once. They should have been taken earlier. To press vigorously on is one of the best ways in which we can fortify ourselves for the future, to build a basis on which we can again fight effectively if we must, to furnish a safeguard against having to fight at all. The by-products, in improved health and a higher standard of living, will pay the costs many times over.

CHAPTER XVII

PLANNING

". . . indeed one of the most important characteristics of the successful officer today is his ability to continue changing his methods, almost even his mental processes, in order to keep abreast of the constant change that modern science, working under the compelling urge of national self-preservation, brings to the battlefield."   —DWIGHT D. EISENHOWER
Crusade in Europe. 1948

A GREAT DEAL is heard about planning these days, and the word has been abused. At times it does not mean planning at all, but the detailed execution of plans by a horde of bureaucrats who, so those to whom the word is anathema tell us, would dictate to us all how many chickens we could have in the back yard and ultimately what to read and how to part our hair. We want nothing of detailed regimentation in this country; our exuberance in throwing off wartime controls shows this. But we most certainly do want sound planning for the future, and we expect our legislature to regulate our economy soundly in accordance therewith. We want our planning done by sound men in whom we have confidence, and not by faddists, advocates of a rigidly socialized state, or advocates of any ism. Moreover, we want the kind of planning that will release the energies of our people so that the competitive free-enterprise system can work still better. If we stick to the dictionary, the word "planning" still means bringing the light of reason to bear on the future as a basis for logical action.

No argument is needed to demonstrate that it is essential, for the national safety, that we do military planning well in these days of uneasy peace. This can be no analysis of the organizational system under which it is done or the law that controls it. Such a treatment would require a book by itself.

But we can, on the basis of this review of the nature of war
and of democracy and of the way in which science interpene-
trates and influences both, bring out some of the factors involved, not to prescribe a new system but rather in an attempt
to see more clearly why the present system fails.

We are confronted immediately by an apparent paradox. In
recent times, we have done military planning of actual campaigns in time of war exceedingly well, and we have done
military planning of broad nature in time of peace exceedingly
badly. Yet both have been done largely by the same individuals, and so it will pay to look for reasons.

It requires little demonstration to show this contrast, and
one example of the planning of a campaign will suffice. The
invasion and conquest of the continent of Europe, beginning
with an amphibious operation in the face of an able and active
enemy, was undoubtedly the greatest and most complex military operation of history, involving great numbers of men and
vehicles, enormous masses of highly diversified supplies, and
the utilization on a wide scale of novel and intricate techniques. It was also, without much doubt, the best planned and
executed large military operation that was ever undertaken.
The record is complete, recent, and fully before us. It shows a
co-ordination of the operation of allies, smoothly and effectively as a team, to an astounding degree. The timing was magnificent; the elements of surprise and deception were fully utilized; the decision to move at the proper moment in spite of
the threat of weather was one of the momentous decisions of
all time. We can take pride that the allied democracies could
learn to work together well enough to accomplish it, that they
could produce the great military leaders able to carry it out,
and that they had the sense to give them the support and backing essential to the performance of their task. It should dispose of the idea, for a long time to come, that, when it comes
to tackling arduous and hazardous problems in a comprehensive and intelligent manner, there is anything seriously wrong
with the military mind or the military life that molds it. Our

planning in peace may be bad, but not for any such cause.

That our military planning in peace is faulty also needs little demonstration, for we have the experience of the postwar years fresh before us. How have we determined such vital questions as the fraction of our effort to be placed in strategic air facilities, or whether an outsize aircraft carrier is now worth its great cost? By careful judgment in which expert opinions are balanced, supplemented, and vitalized by coolheaded public discussion? No. Rather, by arguments of these highly technical matters in public, in the press, in magazine articles, some of them vitriolic and most of them superficial. By statements of high-ranking generals and admirals attacking one another's reasoning, and at times almost one another's veracity. By presidential and Congressional commissions paralleling almost entirely the organization for planning purposes established by law. By the action of committees of Congress, based on superficial examination of the facts and analyses, attempting to pick out from the chaos something that corresponds to reason. By the personality and appeal of enthusiasts for this or that, wherever placed. This is not planning; it is a grab bag. It will lead us to waste our substance. It will lead to strife between services of a nature that can destroy public confidence. It will render us vulnerable in a hostile world. It has already done so to an intolerable degree.

Why the striking contrast? There appear to be at least three reasons. First, peacetime planning deals with the facilities and techniques of the future rather than the present. Second, the bond that holds men in unison under the stress of war becomes largely dissolved when peace returns. Third, peacetime planning is done in a political atmosphere and arena.

The first reason is concerned with industrial capacity, suppliers of strategic materials, contractual relations with private industry, and a maze of similar matters. It is better exemplified, however, in the scientific field. It is a far easier matter to grasp the performance and usefulness of a novel device actually ready and at hand than to understand the trends of science and the potential influence upon warfare of their future applications. The behavior of men in battle can be soundly estimated only by those who have spent a lifetime in military affairs; the applications that will probably flow out of future science can be soundly estimated only by those who have spent a lifetime developing and utilizing science. Military men have arrived rather generally at the first stage, where they can grasp the value of a device before them; they have by no means arrived at the second, where they can visualize intelligently the devices of the future. Yet military planning for the future that ignores or misinterprets scientific trends is planning in a vacuum. Military men are therefore in a quandary; there is a new and essential element in their planning that they do not understand. To leave it out is obviously absurd. To master it absolutely is impossible.

Much the same situation applies to other areas: psychological warfare, civilian defense, protection against subversive attack. To leave all of these things out of consideration would be to move military planning into a corner. Suddenly to become recognized authorities in all such fields is just too much for the professional military mind; it would be too much for any single professional group whatever.

We have arrived at the point where military planning of adequate comprehensiveness is beyond the capacity of military men alone. Either they will learn to cope with the new situation or they will lose their franchise. So far, the course of events has come close to the latter alternative. The days are gone when military men could sit on a pedestal, receive the advice of professional groups in neighboring fields who were maintained in a subordinate or tributary position, accept or reject such advice at will, discount its importance as they saw fit, and speak with omniscience on the overall conduct of war. For one thing, professional men in neighboring fields have no present intention of kowtowing to any military hierarchy, in a world where they know that other professional subjects are just as important in determining the course of future events in the nation's defense as are narrowly limited military considerations.

There is a solution to this quandary, of course: it lies in professional partnership. Specifically, it lies in a system by which a central professional military planning group will take the lead and the ultimate responsibility under conditions where they can avail themselves of the collaboration of the best minds that the country produces in all neighboring professional fields, in physics, chemistry, biology, mathematics, engineering, medicine, psychology, statistics, law, organizational theory and practice, and other fields that now enter intimately into the whole. If military men attempt to absorb or dominate the outstanding exponents in these fields, they will simply be left with second-raters and the mediocre. To avoid this they must accept at the outset the principle that in the process of synthesizing the judgments of diverse specialists into the integrated whole of a comprehensive plan, they will not override the professional judgments of others within the areas where those others have special competence. There is no doubt that professional men of eminence will serve assiduously and patriotically in such a system, if the conditions of service are genuinely dignified and constructive, but there is also no doubt that they have no intention of being pushed around or being placed in an inferior status, or of placing the judgments at which they arrive by the sweat of their brows before men of another profession for inexpert dissection or distortion. The professional men of the country will work cordially and seriously in professional partnership with the military; they will not become subservient to them; and the military cannot do their full present job without them.

There has been very significant progress in this direction since the war, especially in the scientific field. Men of stature in military circles see the situation clearly. But it is by no means yet on a sound basis, nor has it proceeded nearly far enough. Before it can become a reality those military men of vision who grasp the true relationship must combine and work the system out and, incidentally, see to it that little military men, who would manipulate all civilians into subordination or exclusion, are placed where the harm they can do will affect something less vital.

Military men will have to do the job themselves; it cannot be imposed from above. We have a system in this democracy that places the military subordinate to civilian political authority; the President of the United States is the Commander in Chief of the armed services. This is honestly subscribed to and supported by all broad-gauge military men, and there are surprisingly few who inwardly dissent. The civilian authority is supreme; it can impose its will by direct orders and by control of funds. In the field of military planning it can lead the horse to water but it cannot make him drink. It could take away from the military the essential job of military planning in a modern sense—it has begun already to do that very thing—and it would be exceedingly unwise to do so completely. It can, on the other hand, provide facilities, funds, and organizational means by the use of which military men can institute a system of planning that does more than nibble around the edges of the actual problem. But civilian control cannot force the military to do the job. This can occur only if military men of high caliber rise to the occasion. They have the capacity and the understanding, if they have the will.

In the meantime we have the sorry spectacle of military men battling one another in public, in a scramble for funds, and it is about time we cut it out of our pattern of national life. It is undignified, immature, disruptive, and damaging to morale and to the country's safety. The public is weary of the feuds, and somewhat disgusted. Why do we still have it, at a time when it is crystal clear that the missions of the services in war are inextricably linked, that we no longer have three separate military arms but a single military system, and that it is essential for our preservation that the system operate smoothly as a unit for its intended purpose?

This time it is not the fault of the military; it is distinctly the fault of the superimposed civilian political authority. There is one very refreshing attribute of military men—they follow orders. When there is a clearly constituted line of command, and an individual who has the authority and the responsibility of making a decision, military men will argue before him with vigor, but when the decision is made they will loyally abide by it. There is here a striking contrast, as there should be, with college faculties, who usually take a decision as the starting gun for an argument. But there is also a converse attribute. When lines of authority are not clear, or when they are divided and confused, military men will advocate the things they believe in equally vigorously and all over the shop. They will, and they do, and out of this come the public clamor and the emotional controversy that we deprecate. When peace comes, the bonds of intense common purpose tend to dissolve, and men who would agree under stress in order to get on with the job argue interminably instead. We need within our single military system a well-conceived provision for deliberately arriving at all necessary decisions in an orderly manner, and for carrying them out without equivocation or evasion. We need this provision even more in peace than in war. We do not have it fully yet, even though recent steps in this direction have been salutary, and we are suffering the consequences.

There need be no question involved of suppression of honest opinion when this situation is cured. There is no suppression of argument when a court prevents the litigants from arguing a case in the press or publicly criticizing the procedure of the bench during a trial. It is merely a question of ensuring that the arguments are presented in orderly fashion before a tribunal competent to judge them, with all the facts and expert testimony available, and after sober deliberation. The President, or his Secretary of Defense, or a duly authorized committee of Congress, can examine the system and its functioning at any time, and if there is suppression they can correct the situation. If those who sit in judgment are inconsistent or biased they can be removed. If the machinery tends to become bogged down they can set deadline dates and speed it up. The trouble is not that any such system is faulty in conception; the trouble is that the system has not been set up and made to work; it does not yet exist. We do not yet have clearcut lines of command in our defense establishment, extending all the way from the Commander in Chief to the soldier in the ranks, without ambiguity or bifurcation, so that every individual knows his authority and duty and exactly to whom he reports and is responsible. Until we do, we shall have men kicking over the traces, evasion of control, public controversy, lack of internal discipline, and general confusion. For this we may take to task those whom we have elected to operate the government of the United States and make our laws. They have done part of the job of giving us a streamlined, efficient, unified military organization, but not all of the job as yet.

This leads directly to the third point of the discussion. Our military planning in peacetime is conducted in a political atmosphere. Certainly it is conducted in a political atmosphere; what is wrong with that? This is a democracy, not a dictatorship, and the essence of democracy is freedom, the responsiveness of government to public opinion, and uncoerced criticism. In time of war, we place absolute power over important matters in the hands of individuals—the President or a military leader in the field—and by hard experience we have learned to do so and to back them up. In time of peace, we have no intention of doing any such thing; we propose to examine and review, we propose to subject great issues to discussion, we propose that even our greatest men shall run the gantlet of criticism, and we are dead right. Yet there is no negation here of maintaining an organizational system that can examine intricate problems beyond the grasp of the casual layman, in orderly fashion, to produce results in which we can have confidence.

It is done all the time. The Congress sets up and maintains the Federal judicial system. That system is utterly dependent upon Congress for appropriations to maintain it, and it is dependent upon the police power in the hands of the President for the enforcement of its decrees. Yet neither the President nor Congress would think of overturning the judgment of a Federal court, no matter how much they might dislike or resent it, by substituting their own. We have a Federal Trade Commission, to which Congress has delegated the job of judging involved business questions, a Securities and Exchange Commission charged with the duty of regulating the stock exchanges, and many others. As a result of the actions of these, the facts they produce and their judgment regarding them, the President may propose new law and Congress may enact it. The progress of the commissions is under review often, and, if they boldly exceed their authority, they will be brought up with a round turn. If they are incompetent and render foolish judgments, their personnel will be altered; someone will be fired, within the safeguards that Congress itself has established in order to ensure reasonable independence and continuity. But neither the President nor Congress will interfere with their handling of specialized problems as long as they seem to be doing a good job. Similarly, we now need an adequate organization within the National Military Establishment for deliberate military planning in all aspects of modern war, not merely the strictly military part of a generation ago. When the President and Congress have set it up and started it in operation, they should review it constantly, they should review its methods and its personnel, and they must review its decisions, for upon them enormous appropriations will depend. But they should not substitute their own judgment for the judgment of professional men operating in a highly involved area of techniques and conditions. They must lay down for the system the limits within which it functions, financial and otherwise. They must specify, through adequate organization in the State Department and elsewhere, the international political considerations that furnish the timetable and the objectives for military planning. They must integrate the overall military conclusions with the greater problem of maintaining the economic health of the country. But if they set up the system, and then destroy it by becoming military planners themselves, we do not have a logical system for getting sound results, we have a game and a lottery, and the condition of the world is far too serious to tolerate any such luxury or diversion.

There are military experts in Congress. Certainly there are, just as there are many experts in law, a few professors, and even a few highly competent engineers. There may even be a scientist there. These men are particularly valuable; and those who have had military experience and who are close students of military history, strategy, or tactics are especially valuable in the examination of whether the military organization is running well. Yet if they substituted their judgment for that of the organization set up to produce deliberate judgments, we would be playing hunches in a hazardous manner. Sometimes, in the recent past, they have had to—for the system has been faulty and has not produced mature judgments at all, based on all the facts and analysis and the testimony of men of experience and competence in diverse professional fields; it has produced merely controversy and playing to the galleries, and someone had to make a decision and get action. Still, when the system is running in complete form, it is not only these members of Congress of special military background who can judge whether it is operating well for its intended purposes, and who can move to repair it if it spouts oil out of its exhaust. The President, through his subordinates to whom he delegates authority, will have the primary duty of seeing to it that the system is well manned and performs well. But Congress can look in and check, as indeed is its prerogative and duty, and every member of Congress who is charged with so doing by his assignments can participate, and he does not need to be a specialist in order to do so.

It will be well to pause and examine this last point further, for it is at the heart of the success of much of our democratic process. How does it happen that a Congressman who was a small-town lawyer in private life can sit on a committee and judge wisely whether the military organization is running well, whether its proposals appear sound, and whether its appropriations should be cut or extended?

Congress is composed of successful politicians. To be successful as a politician a man needs many talents and a bit of luck or misfortune as well, according to the point of view. Especially he needs one attribute: the ability to judge men, the ability above all to know instinctively whom he can trust. Without that quality he does not get to Congress. With it, in Congress, he performs most of his functions. Without it, in the maze of technicalities and specialized legislation, he would be utterly lost. How did Congress decide during the war to throw billions into the race for the atomic bomb? Because there were leaders of Congress who had the confidence of their fellows and because those leaders trusted the elder statesman who was then Secretary of War, Henry L. Stimson. All that they cared to know about the bomb was explained to these leaders freely, the status of the program and the plans, the nature of the results that could flow out of it, the bizarre techniques involved, the progress of enemies as far as it was known. Did they understand the whole subject of atomic energy? Not by a jugful! They understood men and their relations to organizations, and they had the patriotism and determination to join in a tough gamble when it was headed by a man of unquestioned integrity, understanding, and courage, and when they saw nothing to arouse their qualms in the men that were backing him up. Why did President Roosevelt back the program wholeheartedly from the outset? For exactly the same reason. It is a sound political system that projects into power primarily men who know men. When they are also men of high patriotism, as most of them inevitably and fortunately are, then, in spite of shenanigans, political maneuverings of questionable nature, occasional blatherskites, and sometimes profound ignorance, they can create and cause to function well all sorts of organizations that can deal effectively with the twisted strands of present-day science, technology, economics, and military art, without understanding any one of these in its broad compass and without interfering improperly with the activities of those who do. They can create, supervise, and exercise overall judgment, because they understand men.

This is democracy at its best. We do not often see it at its best,

we see a muddle, and cross-purposes, and small men on pinna-
cles, and all the tinsel and ballyhoo of a political merry-go-round.
We see the reputations of sincere men smeared, and we believe
there are hypocrisy and evil action, nepotism, and enrichment
from the public purse that are never uncovered. We hear raucous
cries and arguments that make no sense, and witness the hard
labor of devoted men disregarded and the false front maintained
in power by intrigue. We see a maelstrom in which men struggle
and contend by methods of elementary barbarity. We see these
things all exaggerated by the more lurid press and columnists,
for a fight is news and harmony is not. Yet, if we see only these
things we do not see democracy as it exists.

To see more we must look closely indeed, through the murk
and beyond the obvious. What we then see is a group of men—
professional politicians—who have had the hardihood and tough
hide necessary to climb the political ladder, and who have maintained a simple honesty in the process, of all shades of political
opinion and all sorts of philosophies regarding the destiny of the
race, of all religions and of no religion, having a pride of accomplishment in representing multitudes, bound together by a common love of country and a common faith in the democratic
process. Some of them lead and many of them follow, but it is
their combined judgment that orders the course of events. We
see an honorable profession, of those who govern men by the
utilization of the machinery of democracy, not by dictation or
intrigue, but by their keen understanding of the nature and
motives of their fellow men and by their sound judgment as to
those in whom they would place their confidence. This is an
amorphous profession composed of men of good will in public
affairs. It has no artificial solidarity emphasized by titles or
symbols; it is bound together only by the common determination
to make democracy work. It can bring order and reason out of
chaos in our military planning. Our future lies in the hands of
this group of men; they will determine how all the rest of us
function and how strong we may be. If they are numerous, determined, and united in the common purpose of preserving our

freedom in a hazardous world, if we have sense enough in our peril to elect and support them, the future is safe in their hands.

For military planning in the present sort of world we need an able group of military men, within the National Military Establishment, capable of planning in the broadest military sense, looking into the future at scientific trends, and dealing with unconventional as well as conventional warfare. To accomplish this they will need to take in, as associates and partners, professional men from many diverse fields, retaining the responsibility for overall planning, but utilizing other brains than their own in the process. The organization should be such that controversy will be settled internally, and for this the lines of authority need to be fully clear and vigorously enforced. If it will, Congress can give us such a system and can make it work. No small part of the task of making it work lies in first making sure it is sound and then depending on it and backing it up.

CHAPTER XVIII

CONCLUSION

"Those who read this book will mostly be younger than I, men of the generations who must bear the active part in the work ahead. Let them learn from our adventures what they can. Let them charge us with our failures and do better in their turn. But let them not turn aside from what they have to do, nor think that criticism excuses inaction. Let them have hope, and virtue, and let them believe in mankind and its future, for there is good as well as evil, and the man who tries to work for the good, believing in its eventual victory, while he may suffer setback and even disaster, will never know defeat. The only deadly sin I know is cynicism."   —HENRY L. STIMSON. 1947

IF WE LOSE our liberties it will be because we abandon them.

It is a hazardous world; it has always been a hazardous world, beset by the perils of harsh nature and the greater perils of harsh men. Ambition and cunning and the ignorance of multitudes have created rigid systems that have suppressed all liberties, and from these men have broken away into freedom at times, have become confused in their councils, and have again succumbed. This time there is hope, for free men have at last created a democracy more effective, as long as it retains its hallmark, than any dictatorship can ever be in dealing with the intricacies of civilization.

There have been recurrent wars, and these have harassed men in their progress toward health, the control of the forces of nature, and the blessings of saved wealth, harassed but not halted advance toward better material things. They have burdened the spirit of man and kept his eyes on the mud about him when he might perhaps have lifted them to the heights.

Now comes the application of science, and it renders war more swift and more rapidly destructive of life and goods. It calls, as never before, for intelligence and common action among whole peoples for its prosecution.

There need be no more great wars; yet there may be. If democracy enhances its latent strength, and free men join in a common purpose, resisting the temptations of avarice and the diversions of petty causes, they can prevent great wars. They can finally mold the whole earth in their pattern of freedom and create one world under law. If democracy loses its touch, then no great war will be needed in order to overwhelm it. If it keeps and enhances its strength, no great war need come again. Yet there is chance and change, a great war may come in ways we do not see, and free men must be ready.

Still, the specter of war should not paralyze as does the glare of the cobra. To intrigue and deception there is now added the weapon of terror in the armament of those who would dominate. All these weapons must be resisted in the uneasy days of peace. Strength cannot be built in unreasoning fear, and strength is essential to prevent a holocaust. Fear cannot be banished, but it can be calm and without panic, and it can be mitigated by reason and evaluation. A new great war would not end the progress of civilization, even in the days of the riven atom, even with the threat of disease marshaled for conquest. It is even possible that defenses may become tightened, not made absolute, but competent to halt the full flood of death from the air. As science goes forward it distributes its uses both to those who destroy and to those who preserve. A great war would be terrible; it would not utterly destroy. It need not destroy democracy, for the organization of free men tends to become refined under stress, whether the stress be hot or cold, and meets its greatest hazards when the times are soft. This has always been true since men have begun to learn their strength in freedom.

The course of history is determined by the faith that men are guided by. If they misread the lessons of expanding knowledge and in their brazen egotism believe that all things are known or knowable, then they will see nothing but an endlessly repeating pattern of sordid strife, the ascendancy of ruthlessness and cunning, man damned to exist a little time on an earth where there is nothing higher than to seize and kill and dominate. If they see beyond this they will see by faith, and not by reading instruments or combining numbers. They may look beyond by religious faith, or they may look merely because they feel validity in the heart's desire and conviction that good will is not a delusion. If they have faith they will build, and they will grow strong that their buildings may endure.

Their greatest buildings will be those of relations between man and man, systems and organizations and law. If they build well the structure will preserve the resourcefulness and initiative of freedom, and further the urge to create, with no stifling regimentation or deadening mediocrity. They will build so that the ambition of youth may have an outlet without artificial barriers, so that genius may rise and innovate for the benefit of all. They will build their structure so as to marshal their full strength in powerful array, designed to deal with assault of any form, whether it come in the full light or skulk in the dark.

If they build thus, and keep the faith, no power on earth can destroy what they create.

INDEX

272

A NOTE ABOUT THE AUTHOR

Dr. Vannevar Bush has been President of the Carnegie Institution of Washington for more than ten years. His early work was in applied mathematics and electrical engineering. Before becoming head of the Carnegie Institution, which is the operating agency for the scientific research work endowed by the late Andrew Carnegie, he was Vice-President and Dean of Engineering at the Massachusetts Institute of Technology. His governmental career began in 1938, when he was appointed a member of the National Advisory Committee on Aeronautics. In 1939, he became its chairman. Before then, his own work in research on weapons had been principally in the field of ballistics and in the detection of submarines.

In 1940 he was named by President Roosevelt to be chairman of the National Defense Research Committee, "to co-ordinate, supervise, and conduct scientific research on the problems underlying the development, production, and use of mechanisms and devices of warfare, except scientific research on the problems of flight." This was later enlarged into the Office of Scientific Research and Development, which included research on military medicine, and Dr. Bush was its director. Also in 1940 the Advisory Committee on Uranium—out of which grew the Manhattan District project and the atomic bomb—was placed under Dr. Bush's direction.

In no other country has a scientist ever been given the wartime powers or the funds which were at Dr. Bush's disposal. Robert E. Sherwood, biographer of the late Harry Hopkins, has called him "an ideal leader of American scientists in time of war . . . his analysis of a tangled situation and his forceful presentation of a course of action produced results far removed from his official sphere of influence."